Praise for *Deep Survival*

"Stunning. . . . Expect to read some hair-raising stories along with a practical guide to survival in the wild."   —*Scouting*

"Riveting accounts of avalanches, mountain accidents, sailors lost at sea, and the man-made hell of 9/11."   —*Sports Illustrated*

"Great stories of disaster and survival where one irresistibly wonders, 'How would I do in this circumstance?' combined with revealing science about the physiology and psychology of how we deal with crisis. [Gonzales's] science is accurate, accessible, up-to-date, and insightful. An extremely good book."
      —Robert M. Sapolsky, author of *Why Zebras Don't Get Ulcers*

"A feast of excitement and wonder. Makes complexity and chaos come alive, girdled by neurological processes, drenched with fantastic accounts of danger and death. You will see the world differently."
—Charles Perrow, author of *The Next Catastrophe: Reducing Our Vulnerabilities to Natural, Industrial, and Terrorist Disasters* and professor of sociology emeritus, Yale University

"A gripping and thoughtful investigation of the greatest adventure of all—survival. Through riveting tales of disaster and endurance, Gonzales explores the icy mental clarity that characterizes survivors."
      —Jerry Kobalenko, author of *The Horizontal Everest: Extreme Journeys on Ellesmere Island*

"Gonzales has masterfully woven together personal survival stories with the study of human perception to reach rock-bottom truths about how to live with risk."
      —Peter Stark, author of *Last Breath: The Limits of Adventure*

"Professional rescuers will love *Deep Survival*. It goes to the heart of the instincts that drive us to risk our own lives to save others."
—Jacki Golike, former executive director,
National Association for Search and Rescue

"*Deep Survival* provides a new lens for looking at survival, risk taking, and life itself. Gonzales takes the reader on a roller-coaster ride that ends with rules of survival we can all stand to learn. Equally important, he answers the question: what is the value of taking risks? I love this book."
—Jed Williamson, former editor of
Accidents in North American Mountaineering

"*Deep Survival* is by far the best book on the many insights into epic survival stories I have ever read."
—Daryl Miller, former chief mountaineering ranger,
Denali National Park & Preserve

"Should be required reading for anyone venturing off the beaten path."
—Jeff Randall, Randall's Adventure
Training School of Survival

DEEP SURVIVAL

Also by Laurence Gonzales

DEEP SURVIVAL

Who Lives, Who Dies, and Why

TRUE STORIES OF MIRACULOUS ENDURANCE AND SUDDEN DEATH

Laurence Gonzales

W. W. NORTON & COMPANY

INDEPENDENT PUBLISHERS SINCE 1923

NEW YORK LONDON

For my father

And you, my father, there on the sad height,
Curse, bless, me now with your fierce tears, I pray.
—Dylan Thomas

The Library of Congress has cataloged an earlier edition as follows:
Gonzales, Laurence, 1947–
    Deep survival : who lives, who dies, and why : true stories of miraculous endurance and sudden death / Laurence Gonzales.—1st ed.
    p. cm.
Includes bibliographical references and index.
  ISBN 0-393-05276-1 (hardcover)
1. Wilderness survival—Case studies. I. Title.
  GV200.5.G66   2003
  613.6'9—dc21

                                                                      2003010867

ISBN 978-0-393-35371-6 pbk.

W. W. Norton & Company, Inc., 500 Fifth Avenue, New York, N.Y. 10110
www.wwnorton.com

W. W. Norton & Company Ltd., 15 Carlisle Street, London W1D 3BS

3   4   5   6   7   8   9   0

CONTENTS

INTRODUCTION TO THE NEW EDITION

I ONCE WORKED with the Chicago Fire Department. I wasn't a fireman, as I had wanted to be when I was four years old. Rather I was a journalist on assignment to tell what it was like in one of the most dangerous professions. Those brave and big-hearted fire-fighters took me in and fed me firehouse lunches of corned beef and cabbage and potatoes. One day they honored me by giving me turnout gear of my own: the Kevlar coat, the thunderous boots, the battered hat with the half-melted faceplate—this gear had been to the inferno, and I respected it, even feeling guilty as I wore it. After all, if I was wearing it, where was its owner? I turned out at incidents of stranded window washers and car wrecks and trash fires. Mostly I turned out at false alarms.

Then came the day when we arrived at a fire, and instead of having me watch from the truck, they led me right into the belly of the beast and showed me how a real fire is fought. It was terrify-ing work, but I was sucked in by its romance. These were the real guys I had dreamed about as a kid. And I knew nothing with more certainty than this: If they got out, I'd get out. I knew that they would not leave me for dead, even if their own lives were on the line.

One day we were working a four-bedroom wooden house. The whole structure was in flames, and I couldn't believe we were going in. We climbed a staircase that was burning beneath our feet. On the second floor, the rear wall of the house had fallen out and we could see the backyard aflame with discarded truck tires and children's play equipment as we fought from our precarious fiery perch. This experience forged a special place in my heart for firefighters.

When *Deep Survival* came out, I began getting calls from firefighting organizations, asking if I would come and share some ideas that would possibly increase their likelihood of survival. I was humbled and excited that firefighters would want to hear from me. I was even more surprised to learn that firefighters across the country and up into Canada were using the book in their training. I have been speaking to groups of firefighters ever since.

It made sense to me that people in one of the world's most dangerous professions would want to read *Deep Survival*, but the next round of calls and letters surprised me even more. I received a note from a very well-known investor who was frequently featured in *Fortune* and the *Wall Street Journal*. He invited all of his employees to read *Deep Survival* and wanted them to hear me speak. Could I come to their offices sometime soon?

Would the dangers faced by firefighters have relevance to the challenges faced by stock market investors? When I wrote *Deep Survival*, it was my hope that I could get such a concept across: All accidents are the same. All hazards—physical, economic, or otherwise—share common features. All of our mistakes are from a family of mistakes, and we can learn from them all. And our ultimate survival—in life, in love, in business—evolves by common rules on a shared landscape.

I became fascinated by the broad range of people who responded to the ideas in the book. I had concentrated my research on accidents in the wilderness because they were exciting and well documented. But as the writing of the book unfolded, I had also recognized a universal quality to the lessons I found there. I had high hopes that people in all kinds of pursuits would recognize the same lessons as applicable to their lives and work. I have not been disappointed.

Some of my earliest readers were doctors who treated cancer and wanted to know why some patients would fight and survive, while others gave up and died, even when the diagnosis was the same for both people. I began to see more and more letters from people in the world of business and finance, asking me to come and talk about managing risk and making better decisions. Soon I was talking to the Navy SEALS, Google, the Lawrence Livermore National Laboratory, and the Sloan School of Management at the Massachusetts Institute of Technology—the list goes on.

Some of the most moving responses came from women who had been in terrifying marriages and found in *Deep Survival* the help they needed to get out. Others told of losing a loved one and finding strength in the book. I heard everything from "You saved my life" to "You changed my business plan." A man wrote to tell me how *Deep Survival* had helped him to understand how it was possible that he and his coworkers could build an irreplaceable ten-million-dollar satellite and then drop the completed product on the floor. Not all of the ideas in *Deep Survival* originate with me: many of them go back to ancient philosophers, while some have come from very recent advances in neuroscience. But their broad appeal has meant that there are few walks of life or professions where the book has failed to find an audience.

DEEP SURVIVAL felt very much like a book that I had spent my whole life writing. It took in the adventurous journalist traveling to exotic places, often exposing himself to risk. It took in my love of science, engendered by the influence of my father, a biophysicist as well as a highly decorated World War II pilot. It took in my deep curiosity about human behavior and the functioning of the brain. It cried out to ask (and try to answer) the question: *Why do smart people do such stupid things?*

In the late 1990s, when I was immersed in reading all the latest neuroscience available, I could feel myself reaching a crescendo of readiness to write the book, but I was filled with doubts. Was I correct in my ideas? Would anyone care? I had so much information, so many stories, that I felt confused and agitated as I flew to Spain to visit my daughter, Elena, who was spending a year of college studying there. I brought my notes and a scribbled outline, and she and I pored over it for days in the bars and coffee shops of Seville. She—more organized than I—helped to bring some order to my thoughts. As I flew back over the Atlantic, I saw clearly the ideas I was trying to convey, which are summarized in the twelve rules of adventure at the end of the book.

Arriving home, I wrote the book in a frenzy of typing that felt as if I were taking dictation from a strange voice in the heavens. As the words appeared before me, I felt like a surprised member of the audience, reading them for the first time. I wrote the last chapter, "The Day of the Fall," as if I were the first fish that had managed to gasp and grope its way onto land.

Yet I still had no idea if the book could speak to the people I was trying to reach, those in pain and in need, those who had lost their way and were searching. Writing a book was not like singing an aria on stage, where you can immediately feel the audience's reaction by the applause. Writers write into a silence that can linger for a long time. It wasn't until almost a year later, after much editing and rewriting and rearranging, that I happened to give the finished (and as yet unpublished) book to my editor at *National Geographic Adventure* magazine. When the phone rang a few days later, I was surprised to hear from him so soon. Jim was a great editor, a good friend, and a merciless critic, not given to flattery. I said, "Hey, Jim, what's up?"

He said, "I'm only about halfway through the book, but I had to call you. It's *reaaaally good.*" Jim had never said anything like that to me. I felt almost embarrassed. It was almost too intimate. But I understood: my message had gotten through to him. And that was important to me.

The book has now been translated into Spanish, Italian, Korean, Japanese, and Chinese (twice). With this new edition, and with the continued letters I receive from perfect strangers, I feel that I can exhale a bit and trust that I did not embarrass myself in attempting to capture the strands of human nature that go into making (or avoiding) big mistakes and from which we weave our dramatic journeys of survival. And I feel that I can thank everyone who has already read the book and welcome everyone who is about to. Please write to me and tell me your story. Everybody has one.

Laurence Gonzales
Santa Fe, New Mexico
May 18, 2016
laurencegonzales.com

PROLOGUE

MOST CHILDREN ARE TOLD fantastic stories, which they gradually come to realize are not true. As I grew up, the fantastic stories I'd heard as a young child turned out to be true. The more I learned, the more fantastic and true the stories seemed.

They were unlike the stories other children heard. They were gruesome, improbable, and sad. I didn't repeat them because I thought no one would believe me. They were the stories of a young man falling out of the sky. Unlike Icarus, who had flown too high, he had not flown high enough. At 27,000 feet, his wing was blown off by a German *Flakbatalion*, which was firing 88-millimeter antiaircraft shells over the rail yards outside of Dusseldorf. And unlike Icarus, he's still alive as I write this.

Federico Gonzales, my father, was a First Lieutenant near the end of World War II. He was piloting a B-17 for the Eighth Air Force, when that organization had evolved into a marvelous machine for turning young men into old memories. He was on his twenty-fifth and last mission, which he was eager to complete, because he and his buddy, David Swift, were going to sign up to fly

P-51 Mustang fighter planes, the knights of the sky. My father was like that, despite having been shot down before. He'd enlisted in the last cavalry outfit before the war. He rode horses at a gallop while emptying the clip of his .45 Model 1911-A, reloading while turning to come back and hit the targets again. When the war started, the cavalry was mechanized, and he began searching for the next best thing. He discovered airplanes. He went out for fighters, but they needed bomber pilots, and as his commanding officer told me forty-five years later, "Your dad had a flair for flying on instruments."

When his B-17 was hit on January 23, 1945, he was the lead pilot for one of those enormous air raids that the United States was conducting at the time. The Commandant of the 398th Bomb Group, Colonel Frank Hunter, had asked my father's regular co-pilot to stand down so that he could fly right seat in the lead plane and see the action. The bombers had taken off in great waves of smoke before dawn, formed up, and churned out over the English Channel from Nuthampstead Base.

They'd reached the target area and were on the bomb run when ground fire from the *Flakbatalion* cut the left wing of my father's B-17 in half just inboard of the number one engine. It was rotten luck. During the bomb run, you couldn't take evasive action or the bombs would go astray. Moreover, his was the first plane in the formation, and the hit was the very first firing. It was a mortal wound to the plane and 90 percent fatal to the crew. The blast was deafening, and my father saw immediately that there was going to be no flying out of this. He turned to his boss beside him and said, "Well, I guess this is it."

Then the plane rolled over, ignoring my father's attempts to right it, and began some sort of inverted flat spin. He couldn't tell precisely what sort, for the world had turned into a nasty soup of unfamiliar colors. He gave the bail-out orders through the intercom to the crew, unsure if the thing was even working or had been shot to pieces by the flak. All the lights, horns, and klaxons were going at once as the plane protested with a great crescendo of whines, groans, and the howling noise coming through exploded wind screens. My father looked over at Colonel Hunter and realized that he was already dead, hit by flak or some bit of flying metal from the fractured plane.

Upside-down, spinning, he groped for the parachute beneath his seat. They'd started at 27,000 feet and he had no idea how high they were, but knew he had to get out. The fliers were supposed to wear their parachutes at all times, but the salty old dogs, as my father was then at age twenty-three, kept them under their seats, because the damned things were so uncomfortable to sit on for ten hours. And anyway, the choices they gave you were, as the fliers liked to say, exceedingly butt-puckering, inasmuch as a pilot descending beneath a 40-foot canopy made a great target for sharpshooters. Even the farmers came out to try their hand at bagging an American flier. The women and children would be gathering, too, to collect the bounty from a shattered B-17: nylon, wool, plastic, metal of all sorts, and silk from parachutes and from the escape and evasion maps.

He couldn't reach his parachute with the stupid harness on, so he released it. The centrifugal force slammed him into the instrument panel with such force that it nearly knocked him out. It cut off his oxygen supply, which was fed through a thick rubber tube running up his chest to his face mask. Smashed against the instrument panel, losing altitude he knew not how fast, he reached up with a hand that seemed made of lead now and pulled the face mask off to get a breath of air. He saw Hunter flopped over, hanging helplessly in his harness. He took a breath. Damn. Probably still above 20,000 feet, he thought, and passed out from hypoxia.

While he was out, his aircraft broke in two amidships. On the ground, an old woman, Mrs. Peiffer, saw something amazing: boys falling out of the sky. Of the ten-man crew, only my father survived, and he was severely injured, as might be anticipated in a five-mile fall.

When he awoke, the motion had stopped. He was crumpled and jammed beneath the instrument panel down by the big naked aluminum rudder pedals. He saw sky outside the shattered canopy, a placental overcast from which he'd been born. A man appeared in the broken window frame, standing on the stub of the right wing. He pointed a pistol at my father's head. He was a local man, a German peasant. The idea of killing an American pilot was not an unpopular one in those parts. My father watched with detached curiosity as the man pulled the trigger.

IN 1958, when I was ten years old, I worked in a medical school laboratory at the Houston Medical Center. My father was a biophysicist there. I convinced him to take me to work with him so that I could find out what he did, which he didn't seem able to explain. I'd been after him about it since I was very little, and by the time I was five, I had started to think that he might have been in the slow group at scientist school. All the other fathers could explain what they did. When I was eight, he started taking me to the lab with him after school and on weekends and letting me wash glassware and do other menial jobs. But gradually, he gave me more responsibility. I learned to make microscope slides before I learned how to dance.

One of my earliest jobs in the lab was to take the trash to the incinerator. The trash often consisted of cut-up mice and such things as come out of a biological sciences lab. So I'd lug the trash bags down the vast tiled corridor, which was dimly lit from either side by the glass vitrines in which the demonstration specimens floated in their baths of formalin. There was a human head sliced into half-inch thick slabs, neat as you please. There were many fetuses at various stages of development. And there was one lady, headless, armless, her torso cut in half from the top of her sternum to her crotch. She floated in formalin like a nightmare of Botticelli's Venus about to be born on an ocean wave.

I proceeded to the furnace and cranked the steel handle until the heavy rusted door opened to reveal a roaring orange inferno within. I was just about to toss in the trash bags when I saw a human arm sticking up out of the flames. At first I was shocked, then frightened. Then I realized that, of course, that's where Venus' arms must have gone long ago, along with a lot of other spare parts. And I thought: What the heck am I doing here? I couldn't answer the question then, but I can now: I was chasing my father, trying to get some of that righteous stuff he had. What else does a son do but try to learn from his father?

Since he was a scientist, I grew up believing in science. That meant I had, before I even knew it, already embarked on a search for some universal laws—the Rules of Life.

MY INTEREST in survival began early, when I was a child and learned what my father had done in the war. That he had lived while so many others had died seemed to me to have so much meaning. I heard the stories over and over and could never seem to plumb their mystery. His survival made me believe that he had some special, ineffable quality. I felt urgently that I ought to have it, too.

Gradually, I developed the idea that to survive, you must first be annealed in the fires of peril. Even his everyday life seemed a peril. All around him were the dead, yet he lived on, laughing. Eventually, I went looking for my own brand of peril. I deliberately took risks so that I might survive them. We lived on a bayou in southeast Texas, and from about the time I was seven, it was my private wilderness, with alligators and snapping turtles, rattlesnakes and water moccasins, and strange displaced characters. My Irish Catholic German mother had so many babies—who could keep track of them all? I pretty much ran wild.

When I was in the fourth grade, I began writing about the risks I took. By the time I was in my twenties, I was doing it as a journalist. After thirty years, I realized I'd been writing about survival all along without knowing it. But I'd always come home from a story wondering: Do I have it now? Am I a survivor? Or is there more?

I became a pilot. I began writing about big aviation accidents, that boundary between life and death where my father had made his bones.

With my interest in science, then, I thought there must be some research that could help me to understand the mysteries of survival I'd encountered. I found otherwise rational people doing inexplicable things to get themselves killed—against all advice, against all reason. A perfectly sensible man on a snowmobile is warned not to go up a hill because it will probably produce a fatally large avalanche. He goes up anyway and dies. A firefighter and experienced outdoorsman knows he is going in the wrong direction but persists anyway and winds up profoundly lost in the wilderness. A number of scuba divers are found dead with air in their tanks. They pulled the regulators from their mouths and died. If you had magically transported them to the surface a moment before they removed their regulators and asked them about their impulse, they would have told you that it made no sense: The regulator was necessary for their survival. If you were able to ask them afterward, they would tell you that they didn't intend to take it out. They intended to live.

After reading hundreds of accident reports and writing scores of articles, I began to wonder if there wasn't some mysterious force hidden within us that produces such mad behavior. Most people find it hard to believe that reason doesn't control our actions. We believe in free will and rational behavior. The difficulty with those assumptions comes when we see rational people doing irrational things.

Those who survive are just as baffling. I knew, for example, that an experienced hunter might perish while lost in the woods for a single night, whereas a four-year-old might survive. When five people are set adrift at sea and only two come back, what makes the difference? Who survived Nazi prison camps? Why did Scott's crew perish in Antarctica while, against all odds, Shackleton's crew survived and even thrived in the same circumstances? Why was a seventeen-year-old girl able to walk out of the Peruvian jungle, while the adults who were lost with her sat down and died? It was maddening to find survival so unpredictable, because after all, science seeks predictability. But as I raked the ashes of catastrophe, I began to see the outlines of an explanation.

Most of what I discovered through the years of research and reporting was not new. I acquainted myself with recent research on the way the brain functions, but also with fundamental principles that have been around for centuries—in some cases, thousands of years—as well as with the psychology of risk taking and survival. The principles apply to wilderness survival, but they also apply to any stressful, demanding situation, such as getting through a divorce, losing a job, surviving illness, recovering from an injury, or running a business in a rapidly changing world.

It's easy to imagine that wilderness survival would involve equipment, training, and experience. It turns out that, at the moment of truth, those might be good things to have but they aren't decisive. Those of us who go into the wilderness or seek our thrills in contact with the forces of nature soon learn, in fact, that experience, training, and modern equipment can betray you. The maddening thing for someone with a Western scientific turn of mind is that it's not what's in your pack that separates the quick from the dead. It's not even what's in your mind. Corny as it sounds, it's what's in your heart.

The farther one goes
The less one knows.

—*Tao Te Ching*

How did I get into this and this and how do
I get out of it again, how does it end?

—Søren Kierkegaard

ONE

HOW ACCIDENTS HAPPEN

Of each particular thing, ask:
"What is it in itself, in its own construction?"

—Marcus Aurelius

"LOOK OUT, HERE COMES RAY CHARLES"

IF YOU COULD see adrenaline, then you'd see a great green greasy river of it oozing off the beach at San Diego tonight. You'd see it flowing one hundred miles out toward the stern of the boat—that's what the pilots call it, a boat, despite the fact that it displaces 95,000 tons of water, has a minimum of six thousand people living on board at all times, and is as long as the Empire State Building is tall.

I'm standing with half a dozen sweaty guys on the LSO platform, which at 8 by 8 feet seems very crowded just now. We're steaming into the prevailing wind at "around 30 knots" (the exact speed being classified), and I'm trying not to be jostled toward the 70-foot gulp down to the water. The steel blade of this boat has ripped up the belly of the sea, and I watch for a moment as its curling intestines glisten with moonlight and roll away behind us.

On my left is Mike Yankovich, the landing signal officer (LSO), in his goggles and cranial, his gaze fixed intently about 15 degrees above the horizon. He's got a heavy-looking telephone handset pressed to his left ear, pickle switch held high in his right hand.

It's called the pickle switch because it looks like a large Bakelite kosher pickle with a silver ring enclosing a black trigger. Yankovich has his index finger and thumb poised to press the cut light or wave-off light switches in case he needs to tell the pilot to add power or not to land. The men inadvertently nudge me toward the edge in their enthusiasm to get a look at the F-18 Hornet that's bearing down on us at 150 miles an hour.

A mile out, it doesn't look like much yet, just a black dart, a darker darkness in a sky full of buzz-bomb stars. I know those monster GE engines are burning kerosene faster than a V-2 rocket, but I can't hear them yet. There's just that silent insect shape, unfolding like an origami airplane, a black bat in the bat black night.

I look at the faces around me. Each man has a lump in his cheek from the Tootsie Roll Pops a Marine passed out a few minutes ago. Their white eyes stare intently at the blossoming shape that's chewing up the stars. But they're not staring the way I'm staring. They're different. They're like kids waiting their turn on the roller coaster. And as the plane, 56 feet long, 40 feet wide, heads straight for us, I'm thinking: *We're all going to die.*

The place where that huge machine is meant to land stretches away only a few feet from us. I can see the dashed white foul line shining against the black nonskid deck ("foul" meaning: you step over it, you die). We are standing beside the arrival end of a very short runway built onto the deck of the boat. It stretches away toward the bow at an angle to the keel. The arresting cables, gray and greasy, slither away toward the starboard side. The theory is that the pilot will come in just right and the hook dangling from his tail will catch one of the four wires, which will stop him.

The rest of the deck is a chaos of action as planes refuel and taxi and launch, the A-6s and F-18s and the sexy old Tomcats (last of the stick-and-rudder airplanes), lumbering like slow beasts to the motions of the yellowshirts and the grapes (purple shirts) in their goggles and cranials, who rotate their gauntlet-gloved hands in cryptic signals as the airplanes taxi and queue up for the cat. In the wild deck lights, with the cacophonous metallic music, it has the air of an atavistic ritual with mighty flaming totems.

If I turn around, I can just see the shooter peering out of his bathyscaph bubble in the deck plates in an eerie sulphur light. There goes another one now—*ka-chunk-whoosh!*—in a sleet storm of metal particles and this amazing hissing scream like someone's tearing a hole in hell. Then two angry afterburner eyes seem to hang motionless in the darkness, as the bat shape shinnies up a pigtail of smoke and is gone.

I hear Yankovich through the headphones inside my cranial and turn back to the F-18 bearing down on us. He's speaking over the telephone handset.

The pilot's quaking voice responds, "Three-one-four Hornet b-b-ball, three-point-two."

"Roger ball, wind twenty knots axial."

He's at a quarter mile, a child in a glass bubble, alone in the night, with the dying yellow stars of deck lights below, the cold wind whittling curls of cloud off the cheesy moon, the whistling thunder at his back, as he hurtles toward the heaving sea, straddling two gigantic flamethrowers.

At last we feel the concussion through our feet. The two-wire, that great fat cable, is turned into a singing liquid instrument by the shock, Ravi Shankar meets the Terminator. It catches the plane like a fish, playing it out 200 feet. The plane shudders all over, as the pilot (Del Rio by name—I had seen it painted on his cockpit rail) hangs in his harness in total G-shock for a moment before he can reach up with a hand that seems to weigh 40 pounds and pull the throttle back to idle. Now the yellowshirts wave him toward the huffer cart where the grapes will refuel him.

So that he can go up and do it again.

. . .

DEL RIO'S performance was a perfect act of survival. There he was, safe on the deck of a big boat. He climbed into a machine full of explosive fuel and had himself shot off into the night with a nuclear steam cat. Then, using only his skill and his superior emotional control, he brought himself back by the remarkable performance of catching a wire that he could not see with a hook that he could not see, using cues that made no natural sense, while going 150 miles an hour in the black-ass night.

Most of us will never get into quite the same jam as Del Rio, but every survival situation is the same in its essence, and so there are lessons to be learned tonight. The first lesson is to remain calm, not to panic. Because emotions are called "hot cognitions," this is known as "being cool." "Cool" as a slang expression goes back to the 1800s, but its contemporary sense originated with African American jazz musicians in the 1940s. Jazz was "cool" compared with the hot, emotional bebop it had begun to overshadow. Some researchers suggest that African American jazz musicians refused to let themselves get hot (get angry) in the face of racism. Instead, they remained outwardly calm and channeled emotion into music as a survival strategy in a hostile environment. They turned fear and anger into focus, and "focus" is just a metaphorical way of saying that they were able to concentrate their attention on the matter at hand.

I'd been searching all my life for that state of cool I'd seen my father exhibit, because it had brought him home in one piece. (Well, a lot of pieces, actually, but they'd knitted back together, more or less, by the time I was born.)

Only 10 to 20 percent of untrained people can stay calm and think in the midst of a survival emergency. They are the ones who can perceive their situation clearly; they can plan and take correct action, all of which are key elements of survival. Confronted with a changing environment, they rapidly adapt. Those are the kind of pilots who are supposed to be flying off the deck of the *Carl Vinson* tonight. Getting back onto the deck is the final exam.

I'D SEEN Del Rio earlier when he came in a bit late for the 1800 briefing in Ready Nine, a steel room where we were all slouched in comfortable maroon Naugahyde chairs, trying to look like we weren't scared out of our wits. Every few minutes the catapult shook the whole boat—*ka-chunk-whoosh!*—as if we were taking Exocet missile fire. Nobody even flinched. Yankovich had just begun the briefing for these, his students, when Del Rio walked in, having obviously gotten up from a nap. The side of his face still bore the imprint of the pillow.

"Hey, got a little rack burn there," Yankovich remarked. "Practicing for the luge run?" They call it the luge run because when you're trying to sleep in those tiny racks and the boat is churning along through the waves and planes are exploding off the deck over your head, it feels like the Winter Olympics meets World War III.

Yankovich, a square-jawed, athletic-looking youth with brown hair, green eyes, and a big grin, knew he could tease Del Rio, because in such a place of hypervigilance as this, where nothing, no matter how subtle, went unnoticed, everyone knew, without even having to stop and consider it, that to be able to drop off to sleep two hours before your first night carrier landing was to display a righteous and masterful state of coolness.

I'd gone to stay on the *Carl Vinson* as part of my lifelong fascination with that boundary region between life and death, that place where, to stay alive, you have to remain calm and alert. The reason it's a boundary region is that not everyone can do it. Some fail. Some die.

Shortly before I arrived, one of the pilots was on final, heading toward the deck. He let his descent rate get away from him and got low and slow, and well . . . some would use the term "panic," but that doesn't tell us much. There were plenty of sensory signals screaming at him that he'd better get on the power. (His hand was already on the throttle. All he had to do was move it a few inches.) The LSO had hit the pickle switch, activating those glaring red lights that mean *You are not cleared to land!* The ball, an obvious light in a big Fresnel lens, was right in front of him, telling him he was low. And, of course, the LSO was also yelling in his ear. Somehow none of it got through.

The impact with the tail of the boat cut the plane in two, leaving his WSO (the guy in the rear seat) squashed like a bug on a windshield and sending the pilot skittering across the deck in a shower of sparks, still strapped into his Martin-Baker ejection seat. The pilot lived, and although I'm not sure he got to try that trick again, I'm reasonably certain that he got to have lunch with the captain.

But the most mystifying thing was how he could have kept on coming toward the boat in the face of so much information telling him not to. That was the real boundary I was after: What was he thinking? He was smart, well prepared, and highly trained. Something powerful had blocked it all, and something had forced him to reach for the deck despite all the information he had that it was a bad idea. It reminded me of a lot of accidents in the wilderness and in risky outdoor sports (river running, for example), where people ignore the obvious and do the inexplicable. That was the mystery I'd been trying to unravel.

WHAT THE PILOTS on the *Carl Vinson* know is this: Shit does just happen sometimes, as the bumper sticker says. There are things you can't control, so you'd better know how you're going to react to them. Yankovich explained it to me: "The launch bar breaks. The shuttle goes supersonic and hits the water brake. The water brake turns instantly to steam from all that energy and explodes. Deck plates come flying up, and you fly right through

the deck plates as you take off. So you eject and land on the deck."
That's what's known in fighter pilot parlance as "Not your day."
But there are also the things you can control, and you'd better be
controlling them all the time.

So this is how Yankovich began the 1800 briefing in Ready
Nine on the *Carl Vinson* that night: "It *will* scare the living shit
out of you. If you taxi to the cat and you don't have a knot in your
stomach, there's something wrong. It's like walking into a closet.
You're going to go right off into a black hole. You're sitting there
sucking oxygen, you'd better have a plan. Because if you don't,
you're screwed, and then you're fucked."

We'd all seen the two helicopters orbiting out there (in case
someone went into the water) and the big yellow crane to pick up
planes that got stuck halfway over the side. And those were for the
lucky guys. The first rule is: Face reality. Good survivors aren't
immune to fear. They know what's happening, and it does "scare
the living shit out of" them. It's all a question of what you do next.

The briefing was not about imparting technical knowledge. If
those guys didn't know that stuff already, they wouldn't be sitting
here with their names stenciled on the backs of their chairs (nick-
names, actually: Hairball, Eel, Cracker, Sewdawg, Stubby). Part of
the briefing was to remind them of stuff they knew already, the
way a hymn does in church, but nothing too complex, because in
what psychologists would call their "high state of arousal," nothing too complex was going to get through anyway.

No, the briefing was more about how Yankovich said things,
and how he said them was with a dark, dark humor. It was a little
ritual, in which everyone was reminded how to look death in the
face and still come up with a wry smile. In a true survival situation, you are by definition looking death in the face, and if you
can't find something droll and even something wondrous and
inspiring in it, you are already in a world of hurt.

Al Siebert, a psychologist and author of *The Survivor Personality*, writes that survivors "laugh at threats . . . playing and laugh-

ing go together. Playing keeps the person in contact with what is happening around [him]." To deal with reality you must first recognize it as such.

In keeping with that view, the pilots on the *Carl Vinson* rarely talked earnestly about the risk this close to flight time. They joked about it instead. Because if you let yourself get too serious, you will get too scared, and once that devil is out of the bottle, you're on a runaway horse. Fear is good. Too much fear is not.

Yankovich continued his briefing: "The steam curtain comes up and you lose the yellowshirt for a minute. You'll be a hero real quick if you have the fold handle in the wrong position, so check that. Spread 'em, five potatoes, and you're all set. Okay, wipeout, the engines come up, see that they match. The safety guys jump up and make sure the beer cans are down. Tension signal. Hands you off to the shooter, and then: head back and four G's. Grab the towel rack. Touch the ejection seat handle and make sure you're not sitting on it. If you lose an engine on the cat, stroke the blowers, twelve-to-fourteen-not-to-exceed-sixteen. Rad Alt: You see you're descending, the wiser man will grab the handle."

What the hell did he just say . . . ?

The first time I heard a briefing like that, I was lost. But that's part of the point: only those who get it get it. A nod is as good as a wink to a blind horse. Just for the record, what Yankovich said was that it would be a very bad idea to try to depart with your wings folded up, as they are for taxiing around on the deck. It takes five seconds for them to lock down into place after you move the handle, so you count off as follows: one-potato, two-potato, three-potato . . . Then, after all the technical bits of the launch process have been checked (the wipeout with the stick to make sure your controls are moving freely, checking to see that the engines are both producing the same amount of power, and so on), you're going to hold onto a metal bar known as the towel rack (because that's what it looks like) to keep yourself from being slammed back by the force of the catapult. And just in case that isn't complicated enough, remember that one of your engines could quit, in which case you have to put the other engine into afterburner (known as the blower because it blows) to get enough power to keep going up (but don't overspeed it, those engines are expensive). And since nothing ever works out as planned, check the radar altimeter, which will tell you if you're sinking, in which case wisdom would dictate that you depart the aircraft with some haste.

Of course, it would be unthinkable to talk like that because, for one thing, anybody could understand you. For another, it would be terrifying.

And after all that, there is still the little matter of landing the aircraft, because, as my father used to say, takeoff is optional but landing is mandatory. Yankovich explained the most salient points: "You're at a quarter mile and someone asks you who your mother is: *you don't know.* That's how focused you are. Okay, call the ball. Now it's a knife fight in a phone booth. And remember: full power in the wire. Your IQ rolls back to that of an ape."

It sounds as if he's being a smart-ass (he is), but deep lessons also are there to be teased out like some obscure Talmudic script. Lessons about survival, about what you need to know and what you don't need to know. About the surface of the brain and its deep recesses. About what you know that you don't know you know and about what you don't know that you'd better not think you know.

Call it an ape, call it a horse, as Plato did. Plato understood that emotions could trump reason and that to succeed we have to use the reins of reason on the horse of emotion. That turns out to be remarkably close to what modern research has begun to show us, and it works both ways: The intellect without the emotions is like the jockey without the horse.

My father didn't fly after the war, and he hardly ever talked about it as such, but when he did, I listened. He used to say, "When you walk across the ramp to your airplane, you lose half your IQ." I always wondered what he meant, but instinctively I felt it. When I was a new pilot, I'd get so excited before a flight that I'd get tun-

nel vision. I'd look at a checklist and be unable to read beyond the
first item: Check Master Switch—Off. Sometimes I'd just sit there
in the left seat, hyperventilating. After years of working at it, flying upside down, flying jets and helicopters, and having a few
"confidence builders," I got to the point where nearly every flight
was almost pure joy. I say almost because, even today, there is the
residual anxiety before each flight, the knot in the stomach, that
tells me I'm not a fool, that I know I'm taking a calculated risk in
pitting my skill and control against a complex, tightly coupled,
unstable system with a lot of energy in it. I'll always be the tiny
jockey on a half-ton of hair-trigger muscle. Fear puts me in my
place. It gives me the humility to see things as they are. I get the
same feeling before I go rock climbing or surfing or before I slap
on my snowboard and plunge off into a backcountry wilderness
that could swallow me up and not spit me out again.

So Yankovich was telling his pilots something that was not only
very important to their survival but that is scientifically sound: Be
aware that you're not all there. You are in a profoundly altered
state when it comes to perception, cognition, memory, and emotion. He was trying to keep them calm while letting them face
reality. He'd seen people die. He knew the power of the horse, and
these were his precious jet jockeys.

WHAT YOU really need to know for survival purposes—whether
it's in a jet or in the wilderness—is that the system we call emotion (from the Latin verb *emovere*, "to move away") works powerfully and quickly to motivate behavior. Erich Maria Remarque
described it perfectly in *All Quiet on the Western Front*, in which
he fictionalized his experiences at the front in World War I:

> At the sound of the first droning of the shells we rush back,
> in one part of our being, a thousand years. By the animal
> instinct that is awakened in us we are led and protected. It is not

conscious; it is far quicker, much more sure, less fallible, than consciousness. One cannot explain it. A man is walking along without thought or heed—suddenly he throws himself down on the ground and a storm of fragments flies harmlessly over him—yet he cannot remember either to have heard the shell coming or to have thought of flinging himself down. But had he not abandoned himself to the impulse he would now be a heap of mangled flesh. It is this other, this second sight in us, that has thrown us to the ground and saved us, without our knowing how. If it were not so, there would not be one man alive from Flanders to the Vosges.

Now we can explain it, at least better than we could when Remarque wrote his novel. Emotion is an instinctive response aimed at self-preservation. It involves numerous bodily changes that are preparations for action. The nervous system fires more energetically, the blood changes its chemistry so that it can coagulate more rapidly, muscle tone alters, digestion stops, and various chemicals flood the body to put it in a state of high readiness for whatever needs to be done. All of that happens outside of conscious control. Reason is tentative, slow, and fallible, while emotion is sure, quick, and unhesitating.

The oldest medical and philosophical model, going back to the Greeks, was of a unified organism in which mind was part of and integral to the body. Plato, on the other hand, thought of mind and body as separate, with the soul going on after death. Aristotle brought them back together again. But it seems that people have been struggling with the split for a very long time indeed, probably because they innately feel as if they have minds that are somehow distinct from their bodies. After the Renaissance, a Cartesian model emerged, in which the mind existed alone, had no location, and was completely independent of the body. To the neuroscientist, the brain is no longer seen as separate but is now considered an integral part of the body, no less so than heart, lungs, and liver.

Moreover, many researchers now regard what we experience as mind and consciousness as a side effect (albeit a useful one in evolutionary terms) of the brain's synaptic functioning. Certainly they all agree that the brain is as affected by the body as the body is by the brain. In fact, the brain is created in part by the body (the other main influence being the environment) in the sense that what the brain does or is capable of doing comes from its synaptic connections, and those connections are forged through what the brain comes to know of the body and the environment. Thinking is a bodily function, as are emotions and feelings.

As Antonio R. Damasio points out in his best-selling book on the brain, *Descartes' Error*, "I think, therefore I am" has become "I am, therefore I think." The brain is the only organ that has no clear function. It makes you breathe, but it's not part of the respiratory system. It controls blood pressure and circulation, but it's not part of the circulatory system either. The concept of body has no meaning without the brain and its extensive network of projections that reach to nearly every cell. As an eminent neuroscientist, Damasio is as qualified as anyone to define the brain, and he calls it an "'organ' of information and government." He put the word "organ" in quotes because it's not exactly an organ either.

The information he writes about is of three kinds: information about the environment, information about the body, and information about the good or bad consequences of interactions between the two. The term "government" refers to the fact that the brain's functions are largely regulatory in nature. The brain provides a continuously changing kaleidoscope of images concerning the state of the environment and the state of the body. It receives images from receptors in the body and from the sense organs that take in the outside world. (The images can be smells, sights, sounds, or feelings). At the same time, the brain provides a stream of outputs that shape the body's reactions to the environment and to itself, from adjusting blood pressure to mating. So the brain reads the state of the body and makes fine adjustments, even while it reads the environment and directs the body in reacting to it. In addition, that process continually reshapes the brain by making new connections. All of this is aimed at one thing only: adaptation, which is another word for survival.

The brain does that job mostly through unconscious learning. It learns, or adapts, by strengthening the electrochemical transmissions among neurons and creating new sites at which neurons can communicate with each other. Axons (the fibers that send signals) grow and form new branches and synapses. Memory is the result.

Doing almost anything generates new links among neurons. The process of learning something and the essence of memory has been observed by neuroscientists in the lab: Genes make new proteins in order to store information, and they make new proteins in order to bring that information back as a memory. This process is called "reconsolidation," because, as Joseph LeDoux, a neuroscientist and author of *The Synaptic Self*, put it, "the brain that does the remembering is not the brain that formed the initial memory. In order for the old memory to make sense in the current brain, the memory has to be updated." This is one reason why memory is notoriously faulty.

There is a new split, too, between cognition and emotion. "Cognition" means reason and conscious thought, mediated by language, images, and logical processes. "Emotion" refers to a specific set of bodily changes in reaction to the environment, the body, or to images produced by memory. Cognition is capable of making fine calculations and abstract distinctions. Emotion is capable of producing powerful physical actions.

The human organism, then, is like a jockey on a thoroughbred in the gate. He's a small man and it's a big horse, and if it decides to get excited in that small metal cage, the jockey is going to get mangled, possibly killed. So he takes great care to be gentle. The jockey is reason and the horse is emotion, a complex of systems bred over eons of evolution and shaped by experience, which exist for your survival. They are so powerful, they can make you do things you'd never think to do, and they can allow you to do things you'd never believe yourself capable of doing. The jockey can't win without the horse, and the horse can't race alone. In the gate, they are two, and it's dangerous. But when they run, they are one, and it's positively godly.

The horse can be amazingly strong. On Mother's Day 1999, Sinjin Eberle and his partner, Marc Beverly, were climbing in the Sandia Mountain Wilderness in New Mexico when a rock weighing more than 500 pounds fell on Eberle, pinning him. Beverly watched as Eberle lifted the rock off of himself. Of course, no one can lift a 500-pound rock. Then again, Eberle did it. When I was reporting on airline accidents in the 1980s, an investigator told me of finding dead pilots who had ripped the huge control columns out of jumbo jets while trying to pull up the nose of a crippled plane.

That horse can either work for us or against us. It can win the race or explode in the gate. So it is learning when to soothe and gentle it and when to let it run that marks the winning jockey, the true survivor. And that is what the dark humor of various subcultures is all about: It's about gentling the beast, keeping it cool; and when it's time to run, it's about letting it flow, about having emotion and reason in perfect balance. That's what characterizes elite performers, from Tiger Woods to Neil Armstrong.

There are primary emotions and secondary emotions. Primary emotions are the ones you're born with, such as the drive to obtain food or the reaction of reaching out to grab something if you feel yourself falling. But the emotional system of bodily responses can be hooked up to anything. Remarque's soldiers learned to connect a deeply instinctive emotional response to the whistling of a shell. There were no high-explosive shells when emotion evolved, but it is handily recruited into the task of avoiding them after only a few experiences to make the connection. The connection, once made, is so profound that taking the necessary action requires no thought or will; it works automatically. The proof that it's a secondary and not a primary emotion is that the new recruits didn't have the same reaction, and they died by the score as a result.

Remarque's observation, and the neuroscience that has confirmed it, can illuminate the way accidents happen. If an experienced river runner is pitched into the water, he will turn on his back and float with his toes out of the water, riding on the buoyancy of his life vest. An inexperienced one, like a drowning swimmer, will reach up to wave or try to grab something. Raising his arms causes his feet to sink.

Forty-four-year-old Peter Duffy died on June 16, 1996, while rafting on the Hudson River, and his accident illustrates how important it is not only to control emotions but to develop the appropriate secondary emotions. "He [Duffy] fell into the river," wrote Charlie Walbridge, who publishes *River Safety Report*. "Facing upstream, he attempted to stand, caught his right foot between two rocks, and was pushed under. His life jacket was stripped off, and he was trapped under three feet of water. . . . Foot entrapment rescues are very difficult. You might as well step in front of a speeding car as get your foot caught in a fast moving river. The victim was warned, but failed to follow instructions." Duffy knew, intellectually, what he should have done. But knowing was no match for emotion.

FEAR IS but one emotion. The instinct to reproduce is another, and it initiates a remarkably similar set of visceral responses, though with striking differences involving the sex organs and glands. Anyone who has ever fallen in love, fallen hard, knows what Yankovich means when he says, "Your IQ rolls back to that of an ape." Emotion takes over from the thinking part of the brain, the neocortex, to effect an instinctive set of responses necessary for survival, in this case reproduction.

During a fear reaction, the amygdala (as with most structures in the brain, there are two of them, one in each hemisphere), in concert with numerous other structures in the brain and body, help to trigger a staggeringly complex sequence of events, all aimed at producing a behavior to promote survival; freezing in place, for example, followed by running away. When the reaction begins, neural networks are activated, and numerous chemical compounds are released and moved around in the brain and body. The most well-known among them is the so-called adrenaline rush. Adrenalin is a trade name for epinephrine, and adrenaline is a synonym for it, but neither is used much in scientific circles. Epinephrine and norepinephrine, which come from the adrenal glands, are in a class of compounds called catecholamines, which have a wide range of effects, including constricting blood vessels and exciting or inhibiting the firing of nerve cells and the contraction of smooth muscle fibers. But it is norepinephrine (not adrenaline or epinephrine) that is largely responsible for the jolt you feel in the heart when startled. Cortisol (a steroid), which is released from the adrenal cortex, also amps up fear, among its other effects. The net result of all the chemicals that come streaming through your system once the amygdala has detected danger is that the heart rate rises, breathing speeds up, more sugar is dumped into the metabolic system, and the distribution of oxygen and nutrients shifts so that you have the strength to run or fight. You're on afterburner. The knot in the stomach Yankovich mentioned results from that redistribution (as well as from contractions of the smooth muscle in the stomach), in which the flow of blood to the digestive system is reduced so that it can be used elsewhere to meet the emergency. (Excellent descriptions of this very complex system can be found in Joseph LeDoux's books, *The Emotional Brain* and *The Synaptic Self*. He refers to the amygdala as "the centerpiece of the defense system.")

Evolution took millions of years to come up with emotional responses. It has not yet had time to come up with an appropriate survival response for Navy fighter pilots on quarter-mile final, trying to land a 50,000-pound stovepipe on the heaving deck of a ship.

Peter Duffy's lack of control over his emotional response allowed him to drown himself in the Hudson River. The fighter pilot who slammed into the back of the *Carl Vinson* was the vic-

tim of a similar effect. A secondary emotion got the best of him on
the approach to the boat. For whatever reason, he was not exercising the necessary control, and he let the plane get too low. I know
how it works. I've done it myself. Most pilots have. Fear in the
cockpit, as Yankovich put it, is a knife fight in a phone booth. You
literally have to fight to move your frozen hand to correct the mistake that you see developing before your eyes. You are split.

Many times before, the pilot must have had the sensation of
turn-buckle twisting terror, followed by the cool flood of relief
upon landing. Even as the hormones produced under stress disrupt
perception, thinking, and the formation and retrieval of memories, they set a potentially dangerous trap by exciting the amygdala. They help to dampen explicit (conscious) memory even
while creating and recalling implicit (unconscious) memories with
greater efficiency. As the fear rises, you become more unable to
deal with it because you're not even aware of the learning that's
propelling you. LeDoux refers to this as a "hostile takeover of consciousness by emotion" as the "amygdala comes to dominate working memory." The body knows where safety is, and when you're a
rookie and really afraid, any successful landing carries with it an
explosive, almost orgasmic sense of release. The pilot had developed a powerful secondary emotion, which told him that safety
and even ecstasy could be found on the ground (or the deck) and
that if he could just *get the hell down,* he'd be all right. He had a
true and physical memory of that sensation, which was a powerful
motivator of behavior developed by coupling that experience with
a primary emotional state. He also had an intellectual knowledge
that if you land when you're already low and slow, you might die.
Unfortunately, he had no secondary emotion for that, since he had
no experience of it. It was an abstract idea, forebrain stuff. It could
not compete as a motivator of behavior.

When a pilot hits the "round down," as they call the back of the
boat, it's called a "ramp strike." As one pilot who flew in the war
on Iraq said, "Those are bad and deadly." He explained the way it

happens. The pilot focuses too much on the thing that he feels is most important at that moment: the deck. Home. It's called "spotting the deck," because it breaks up the natural flow of his scan, which ought to include his meatball, line-up, airspeed, altimeter, and angle of attack. Once he fixes on his landing area, he's done for.

The pilot's rising curve of fear went off the charts in one direction, while the rising curve of his motivation toward the deck went off the charts in the other. The jockey lost control of the horse in the gate.

Experienced travelers in the wilderness and people who engage in risky activities understand. In 1910, two British explorers, Apsley Cherry-Garrard and Robert Falcon Scott, set off for the South Pole. Scott died on that expedition. In praising his traveling companions, Cherry-Garrard wrote that they "displayed that quality which is perhaps the only one which may be said with certainty to make for success, self-control." How well you exercise that control often decides the outcome of survival situations. Whether it means making a split-second decision while scuba- or skydiving or keeping your head while stranded in the wilderness, it is the most important skill to take along. And with more and more novices going into the wilderness for fun, the severe penalties that come with a failure of control are becoming evident in the increasing number of search and rescue operations that are launched to save them or recover their bodies.

STRESS RELEASES cortisol into the blood. That steroid invades the hippocampus and interferes with its work. (Long-term stress can kill hippocampal cells.) The amygdala has powerful connections to the sensory cortices, the rhinal cortex, the anterior cingulate, and the ventral prefrontal cortex, which means that the entire memory system, both input and output, are affected. As a result, most people are incapable of performing any but the simplest tasks under stress. They can't remember the most basic things. In addition,

stress (or any strong emotion) erodes the ability to perceive. Cortisol and other hormones released under stress interfere with the
working of the prefrontal cortex. That is where perceptions are
processed and decisions are made. You see less, hear less, miss more
cues from the environment, and make mistakes. Under extreme
stress, the visual field actually narrows. (Police officers who have
been shot report tunnel vision.) Stress causes most people to focus
narrowly on the thing that they consider most important, and it
may be the wrong thing. So while the fighter pilot was fixed on
landing, he very well might not have seen the lights or even heard
the LSO's voice telling him to go around. The organism was doing
what it knew how to do best: escape danger and get to safety as fast
as possible. The rest of the input became irrelevant noise, effi-
ciently screened out by the brain. So he hit the boat.

I did something very like that when I was a new pilot. I was on
approach to landing at my home airport when the controller told
me I was on a collision course with another plane. But I was so
focused, so fearful, that I literally didn't hear him. I heard nothing,
and I didn't even see the plane. He called me on the radio three
times, and fortunately, my friend Jonas, who was sitting beside me,
told me that the controller wanted an immediate right turn. The
task of just getting the hell down had become so important—so
emotionally motivated—that it occupied what neuroscientists call
"working memory" (which in effect means consciousness or attention) to the exclusion of other stimuli. Only because Jonas was so
close to me and could command my attention by punching me in the
arm was he able to break the lock I had put on working memory.

Emotions are survival mechanisms, but they don't always work
for the individual. They work across a large number of trials to
keep the species alive. The individual may live or die, but over a few
million years, more mammals lived than died by letting emotion
take over, and so emotion was selected. For people who are raised in
modern civilization, the wilderness is novel and full of unfamiliar
hazards. To survive in it, the body must learn and adapt.

Although strong emotion can interfere with the ability to reason, emotion is also necessary for both reasoning and learning. Emotion is the source of both success and failure at selecting correct action at the crucial moment. To survive, you must develop secondary emotions that function in a strategic balance with reason.

One way to promote that balance is through humor.

EVERY PURSUIT has its own subculture, from hang gliders and steep creek boaters to cavers and mountain bikers. I love their dark and private humor, those ritual moments of homage to the organism, which return us to a protective state of cool. It unequivocally separates the living from the dead.

When I was fighting fires with the Chicago Fire Department, trying to learn something about how to be cool while going up in flames, I asked one of the men why he became a firefighter. "I like to wreck things," he said. As we smashed windows after putting out a house fire, I believed him, too. We had an old-timer at the firehouse I was working out of, Bernie was his name, who wouldn't even put on his Kevlar turnout coat. He'd fall asleep in the truck on the way to a fire, and when one of us commented on it, Bernie said, "I could sleep with my dick slammed in a door."

Bernie wasn't the only one, either. The guys called the big beer cooler in the kitchen "the baby coffin." They had dozens of names for different types of corpses—"crispy critters," "stinkers," "floaters," "dunkers," and "Headless Horsemen," just to name a few.

Butch Farabee, national emergency services coordinator for the National Park Service, told of taking his friend, Walt Dabney, on his first body recovery in Yosemite (there are a lot of them). They found the man they were looking for, Rick, after he'd been dead a week. "It was just terrible," Farabee said. "His body was quivering with maggots. He was as stiff as a basted turkey, too; we had to break his arms to get him into the body bag. When we lowered the body bag over a cliff, we dropped him. Walt and I had to spend the night out there with the body. I started talking to it, saying, 'Hey, Rick, how's it going today? Sorry about dropping you.' Walt thought I was either terribly disrespectful or out of my gourd. The fact is you have to deal with these things to the best of your ability. If you don't work with it, it'll get you. A dead body is not something you get used to."

Some high-angle rescue workers call body bags "long-term bivvy sacks." It sounds cruel, but survivors laugh and play, and even in the most horrible situations—perhaps especially in those situations—they continue to laugh and play. To deal with reality you must first recognize it as such, and as Siebert and others have pointed out, play puts a person in touch with his environment, while laughter makes the feeling of being threatened manageable.

The grotesque humor of the fighter pilots, then, that secret language, contains truths we don't even know we know. Moods are contagious, and the emotional states involved with smiling, humor, and laughter are among the most contagious of all. Laughter doesn't take conscious thought. It's automatic, and one person laughing or smiling induces the same reaction in others. Laughter stimulates the left prefrontal cortex, an area in the brain that helps us to feel good and to be motivated. That stimulation alleviates anxiety and frustration. There is evidence that laughter can send chemical signals to actively inhibit the firing of nerves in the amygdala, thereby dampening fear. Laughter, then, can help to temper negative emotions. And while all this might seem of purely academic interest, it could prove helpful when your partner breaks his leg at 19,000 feet in a blizzard on a Peruvian mountain.

It is not a lack of fear that separates elite performers from the rest of us. They're afraid, too, but they're not overwhelmed by it. They manage fear. They use it to focus on taking correct action. Mike Tyson's trainer, Cus D'Amato, said, "Fear is like fire. It can cook for you. It can heat your house. Or it can burn you down." And Tyson himself said that fear was "like a snap, a little snap of light I get when I fight. I love that feeling. It makes me feel secure and confident, it suddenly makes everything explosive. It's like: 'Here it comes again. Here's my buddy today.'" It's a dangerous place to be, too. Control can easily slip away, as Tyson's unusual behavior will attest.

I've spent the better part of my life working around people who risk dying a horrible death of their own making. They see it. They're near it. They all have friends who have gone that way. And they all have a strategy for avoiding it—a strange amalgam of superstition, knowledge, illusion, and confidence. But everyone begins with the same machinery, the same basic organism, and when it's threatened, whether in pursuit of pleasure, for duty and honor, or by accident, the organism reacts in predictable ways. It is only by managing and working with those predictable, inborn reactions that you're going to survive. You can't fight them, because they are who you are.

RIGHT BEFORE the planes launched off the *Carl Vinson*, following the 1800 briefing in Ready Nine, I went to dinner with Mike Yankovich and a group of fliers in the officers' mess, ensuring that we'd have that knot in our stomachs. After we'd finished eating, a waiter in a white coat came to the table, and every officer sitting around me said one word to him: "Dog."

When they'd finished, the waiter turned to me and asked, "Dog, sir?"

"Sure," I said. Then, as the waiter left, I asked Mike, "What's dog?"

"Auto-dog," he said. "It's soft-serve ice cream. Like Dairy Queen."

I asked why it was called dog.

"Go over and watch it come out of the machine," he said.

Survival, then, is about being cool. It's about laughing with an attitude of bold humility in the face of something terrifying. It's about knowing the deepest processes of the brain, even if, as nonscientists, we can explain them only through the darkest humor imaginable.

So here they are, these F-18 pilots, about to go up and possibly die doing something horribly risky in the unholy night, and they are joking that for dessert they eat feces.

It's an old habit. Remarque wrote, "We make grim, coarse jests about it, when a man dies, then we say he has nipped off his turd, and so we speak of everything; that keeps us from going mad; as long as we take it that way we maintain our own resistance."

AN HOUR after dinner, I stand on the LSO platform and Yankovich holds the pickle switch high, the heavy telephone handset pressed to his ear. We watch a nervous pilot come wobbling in. I haven't even mentioned the remarkable skill and perception it takes for Yankovich to know, by eyeball alone in the asphalt night, whether or not the black bat we see unfolding before us is going to hit the correct wire. But this pilot's approach looks really bad. Even I can tell.

Through my headphones I hear Yankovich say, "Look out, here comes Ray Charles."

As he releases the pickle switch trigger to send the pilot around for another try, Yankovich does a few dance steps, his head lolling around like a blind man's, reeling there on the tiny LSO platform seven stories above the heaving of the meterless sea.

Yankovich and I turn and watch the jet shoot off the other end of the boat, engines roaring. The plane dips a bit, and we wait until it's securely back in the air. Then Yankovich says to me, "Boy, did you see him *settle*? He'll be picking the seat cushion out of his asshole about now."

TWO

MEMORIES OF
THE FUTURE

A GROUP OF TWENTY snowmobilers had just completed a search and rescue mission to bring out three others, who'd had engine trouble and become stranded overnight at Middle Kootenay Pass in Alberta, Canada. After successfully locating the three, eight of the rescuers sped off to return them to civilization, while the remaining dozen mucked about in the snow with the broken machine. On the way back, an additional fragmentation of that group occurred, because eight of the dozen had been seized by the need for speed and just raced on ahead.

They eventually stopped to wait for their slower companions. They were at an old well site, sitting on their snow machines on a broad, flat area beneath a hill. The hill was well known as a great one for hill climbing, which is also called hammer-heading or high-marking. The idea is to accelerate across the flat terrain and race up the hill as fast as you can until gravity stops you or until you turn back downhill. It's a competitive game to see who can go the highest.

The official report would eventually remind us that "They had

all been specifically told that there was a high avalanche danger .
and that high-marking or 'hammer-heading' was out of the question that day."

But sitting there in the chill air with the view and the spicy
juniper smell of the mountains was intoxicating. The big wide
clearing led up to tremendous vaulted peaks, which seemed to leap
through the gently falling snow and into the low deck of clouds
like a ramp to heaven. The desire to ride a fast, open machine such
as a motorcycle or a snowmobile is evidence of a certain propen-
sity toward sensation seeking, as the psychologists call it. In addition, that particular group had demonstrated poor impulse control
(boldness, or a willingness to take risks, if you like) by racing
ahead of the others. Now there they were, with their throttles in
their fists and all that physical power ready at a touch. There was
more sensory input to urge them on: the throaty animal roar of
the engines. The horsepower throbbing between their thighs. And
there was Mother Nature rising into the gauzy curtains of falling
snow, which both concealed and revealed her great, concupiscent
backside in the swaying fabric of the clouds.

Suddenly, something clicked in one of their brains. The others
watched, startled, as he cut loose, flat out, across the open ground.
He could feel the spinning cleats dig in, as the G's loaded up in his
center of mass. It was a familiar solid feeling of power, as he was
propelled up and up and up the slope, across the 5 inches of new
snow that had fallen in the last two days, which lay on top of 2
inches of rain-saturated snow pack, which in turn sat on another 2
feet of snow that had accumulated in the past two months. All of
that was balanced precariously on top of a weak layer of faceted
wet grains, which are peculiar formations in nature with roughly
the friction coefficient of tiny ball bearings. The whole system was
angled downward at about 35 degrees, which happens to be the
critical angle at which most avalanches occur. On steeper slopes,
the snow tends to slide off before it can consolidate into a slab. On
shallower ones, it remains fairly stable.

The snowmobiler hadn't quite reached the top of the hill when he became bogged down in deep snow. The others below could see him up there like a ballistic nylon bug against the creamy white flank of the mountain and could hear the faint buzzing of his engine as he struggled to get free.

The thrill of the hunt, like so many moods, was contagious. A second snowmobiler goosed his engine, too, and went hell-bent for high ground, right up the crack in Mother Nature's fine white fanny.

It was about twenty minutes to noon, and some of the others must have been wondering: *What are they thinking?*

The second snowmobiler had nearly reached his buddy with this vague, half-formed idea in his head of what he was doing and this little voice, way off, saying, "high avalanche danger" and "hammer-heading is out of the question." It seemed almost as if he had two brains and they were having an argument over his body.

Now those who were still waiting below could read the tally like a great natural scoreboard, the first streak going straight up and ending in a kind of blob, a hole, and the second one growing longer and longer as the roar of the engine dwindled to a faint buzzing, and the black bug grew small against the white slope. Then they heard the rifle-shot crack that typically accompanies the release of a big slab avalanche.

According to the official report: "At about 11:40, Snowmobiler 2 also sped up the slope and when he was about two thirds of the way up a size 3 avalanche released. Snowmobiler 2 was able to ride out to the side of the avalanche and escape. At the bottom of the slope the other six had seen the avalanche start and five of them managed to ride out of the path of the slide." But there was one guy down there who just froze.

The avalanche released a 450-foot-wide swath of snow, 32 inches thick, all the way from the top of the ridge. Once it started, it rushed down 400 yards, cascading on those ball-bearing grains like eight lanes of concrete Interstate highway sloughing off in an old-fashioned San Francisco earthquake.

The first man, the one who became bogged down, was swept all the way down the hill and covered over by 9 feet of snow. But the poor guy down below, the one who froze, he watched, probably mystified, as that wall of snow inundated him in 6 feet of snow like wet concrete. Freezing is a classic emotional response of all mammals. A bystander happened to be videotaping the crowd when a bomb went off at the Olympic games in Atlanta in 1996, and the freezing (and crouching) response of the people is a dramatic illustration of a primary emotion.

The snowmobilers were all wearing transceivers (radio devices meant for locating victims buried by avalanches), so they obviously had some knowledge of the phenomenon, even apart from the warning they'd received. They may, for example, have known that to be buried under even a few inches of snow is a dodgy proposition. Snow is very heavy and sets up in seconds after the avalanche stops moving. The faceted grains refracture as the snow slides and then interlock tightly when they come to rest. The slope in this case dumped 2 million pounds of snow into a terrain trap.

Because they had transceivers, those who were not buried were able to locate the victims immediately, which is a good first step in avalanche rescue. In addition, help from the nearby ski area arrived within minutes at 11:55 A.M. Most victims don't get attention that quickly. But digging 6 to 9 feet of wet snow out of an avalanche deposit, even with everyone there and all the catecholamines pumping, is like digging up a city sidewalk. It took twenty-five minutes to uncover Snowmobiler 3, the one who'd been sitting there flat-footed at the bottom of the hill, gaping up in wonder as the whole mountain came apart before his eyes. He was, of course, blue and not breathing. Snowmobiler 1 wasn't dug out for forty minutes. By then a doctor had arrived to pronounce the two officially dead at two o'clock in the afternoon on the day before Valentine's Day 1994.

The report added: "One of the original search party afterwards questioned, 'Why did they do it? We were told not to.'"

Why, indeed.

As William Faulkner wrote in *Light in August*, "Man knows so little about his fellows. In his eyes all men or women act upon what he believes would motivate him if he were mad enough to do what that other man or woman is doing."

But in the light of recent research, neuroscience can propose a better answer to the question that is asked so often after such accidents: *What were they thinking?*

You have to begin by asking why anyone would want to ride a snowmobile in the first place. It's not as if they had somewhere to go. (They did, but it wasn't up that hill.) You have to ask what could possibly motivate a rational, well-informed person to ride up a slope when he knew that all he could do was come straight back down, having burned through some expensive fossil fuel.

The answer would seem self-evident: It's fun. But then you need to know what makes a completely pointless activity fun. If, as the research tells us, all behavior can be traced to survival strategies, you must ask where the survival value is in that act. We know, for example, why sex is fun: It keeps the species going. If it weren't irresistible, no one would be mad enough to do it.

But modern neuroscience would explain it another way. Like Remarque's soldiers throwing themselves down without thinking at the faintest whistling of an incoming shell, the snowmobilers who went up the hill were acting out a secondary emotion. They had developed it by experiencing it, by synaptic learning. Perhaps it was an emotion that motivates running down prey, a thrilling bodily response that requires speed and swift action to pursue, catch, and kill. Perhaps it was an emotion involved with mating. Perhaps several emotional states were pulled together into a new combination, a heady elixir of chemicals that pour forth when one rides a snowmobile. (Since young, brain-dead, male motorcycle riders supply many of the hearts transplanted in the United States, we know that the same is likely true of that experience.) As generic as emotions tend to be at some levels, the important point is the feeling that resulted and the fact that it was instantly available, without conscious thought.

There's a common confusion about the words "emotion" and "feeling." William James, the father of psychology, was the first to point out that we do not run because we're afraid of bears, we're afraid of bears because we run. The emotion comes first—it's the bodily response (freezing, flight, sexual arousal). The feeling follows (fear, anger, love). The fear associated with being in an earthquake may produce some chemical reactions that are similar to those produced during sexual arousal. But the two experiences are quite different. "The earth moved" can have different meanings in different circumstances. That's why risky behavior can be fun. Fear can be fun. It can make you feel more alive, because it is an integral part of saving your own life. And if the context is one that you perceive as safe, then it's easy to make the decision to take the risk. Your body can make it for you.

But killing yourself is no fun, and so you still have to ask why, when the snowmobilers knew how high the chance of an avalanche was, they decided to take the risk anyway. Even though it is now possible to understand why going up the hill would be fun, it's still not clear how those two who ran up the hill made that decision when they knew it was likely to kill them.

Think of chess masters and how they play. Let's imagine that life is a board game. Some people like to play checkers, some people like to play chess. And those of us who go into the wilderness and engage in dangerous sports are playing chess with Mother Nature.

A computer uses pure logic to play, trying millions of conceivable moves, not every move, but known patterns of moves, put in by chess-playing programmers. People not only don't do that, they can't. In fact, neither people nor computers can play chess by logic alone. There are only a few simple rules to the game but an estimated 10^{120} possible moves in any game of chess. That number is so large that it might as well be infinite. As James Gleick pointed out in his book *Chaos*, there are neither that many elementary particles in the universe nor have there been that many microseconds of time since its creation perhaps 13 billion (1.3×10^9) years ago. Logic doesn't work well for such nonlinear systems as chess and life.

So there they are, in the board game of life, with billions of bits of information stored and new ones coming in fast and furious, and they're at a decision point. How do they decide what to do?

The act of riding a snowmobile up a steep hill had come to elicit an emotional response. Those riders had done it before and had been rewarded with a good feeling. It was a physical feeling, and the body liked it, so it was bookmarked, so to speak. *Note to self: Try hammer-heading again.* The brain creates such bookmarks (technically known as "somatic markers," a term coined by Antonio Damasio) because logic and reason are much too slow if we are to get around in this big old goofy world.

Consider what eating would be like if you used a purely deductive process. Say your brain suddenly turned into a computer and had only logic to work with. It's Sunday afternoon. You're sitting at your desk in your study at home, catching up on some paperwork. You feel hunger. (Here's why computers are so frustrating to people.) First you try eating the telephone. That doesn't taste good, so you try the paperweight. Bad idea. Then you gnaw on the keyboard for a while. Kind of bland. Chew the edge of the desk, the chair, lick the floor . . . Eventually, perhaps a week later, you have gone through the house in search of food, using the system of pure reason, and you reach the refrigerator. After chewing on the handle for a time, you open the door and find that there's some leftover pizza . . . Ah, success!

It sounds ridiculous. You might say, "Of course, I wouldn't try to eat the telephone. I know where the food is." But chimpanzees that have had certain emotional components of their brains removed do something not unlike that. To have reason cut off from the high-speed, jump-cut assistance of emotion is virtually incapacitating, as many neuroscientists have shown with patients who have suffered brain damage. Those patients can perform all sorts of logical functions; they have normal memory; and yet they are incapable of scheduling an appointment because their pure reason makes it impossible for them to decide. They can't bookmark feelings. They have no intuition, no gut feelings. They've been cut off from their bodies in a sense. The most remarkable discovery of modern neuroscience is that the body controls the brain as much as the brain controls the body.

Most decisions are not made using logic, and we all recognize that fact at least at an unconscious level. LeDoux writes, "Unconscious operations of the brain is . . . the rule rather than the exception throughout the evolutionary history of the animal kingdom" and "include almost everything the brain does." In finding food, you have bookmarked a place where the good feeling of satisfying hunger comes out of opening the refrigerator door and grabbing something inside. You remember that the refrigerator is there to hold food, but you don't need to think through it. You can be on automatic pilot, reading something, as you mosey over to the kitchen, open the door, and only then look up to do the conscious part—grab a slice of pizza instead of leftover Thai barbecue. The search for food occurs without anything you'd call cognition, deduction, or logic. Your hunger, your body, leads you there.

When a decision to act must be made instantly, it is made through a system of emotional bookmarks. The emotional system reacts to circumstances, finds bookmarks that flag similar experiences in your past and your response to them, and allows you to recall the feelings, good or bad, of the outcomes of your actions. Those gut feelings give you an instant reading on how to behave. If a previous experience was bad, you avoid that option. When it was good, "it becomes a beacon of incentive," to use Damasio's words. In a similar fashion, the smell of roses can transport me to my grandmother Rosa's house in San Antonio in 1958. You don't have to have sensory input to trigger the effect. You can have an idea or a memory instead.

That instant physical feedback, those feelings that are located through emotional bookmarks, will more or less force a decision unless checked by higher consciousness. It explains why the Navy fighter pilot who was low and slow wound up hitting the stern of the aircraft carrier. He had no idea why he kept coming in the face of clear information telling him not to.

Damasio wrote, "When the bad outcome connected with a given response . . . comes into mind, however fleetingly, you experience an unpleasant gut feeling." Using that system, you can choose very quickly and may be unable to explain your choice afterward. The best and worst decisions are made that way. You don't have to think about it. It just feels right.

The good feeling of riding snowmobiles had become bookmarked for instant reference by the emotional system. That activity, while irrelevant to survival, became as pleasurable as the experience of the primary emotion, because it harnessed the real thing. There are no fake emotions.

Because the system is designed to work without the assistance of logic or reason, there's now an answer to the question: *What were they thinking?* They weren't. The whole point of the system is that you don't have to think.

A CLASSIC experiment was performed in Switzerland in 1911 by a psychologist named Edouard Claparède. Claparède's forty-seven-year-old patient had no short-term memory. One day, Claparède went in and shook hands with her, as he had always done. Only this time, he had a pin in his hand, which stuck her. Although she had no memory of it after a few minutes, she would never shake hands with Claparède again. She had no idea why. She just got a bad feeling when he stuck out his hand. There was some residue of her painful experience that had connected itself to the sight of Claparède sticking out his hand, and she had what, in a normal person, we could call an intuition, a gut feeling. A normal woman couldn't explain those feelings either, but a few milliseconds later she'd remember.

Thinking logically could not assist the patient in deciding whether or not to shake Claparède's hand. But somewhere in her brain and body, her experience had caused her to bookmark the bad feeling of an emotion. She'd involuntarily jerked her hand back when she felt the pain. The feelings that followed the emotional response of jerking her hand back were powerful and unpleasant—shock, surprise, fear. Certainly her heartbeat and breathing sped up. Perhaps her face flushed. She may have cried out. And although she had no conscious recollection of the events or feelings, she was forever after compelled by that emotional bookmark to select the same response. It was a momentous discovery on Claparède's part: his patient could learn without explicit memory or thought. It was as if her body could learn. It hinted at a whole hidden system within her. The very system by which the two snowmobilers decided to rush up the hill against all reason.

On the other hand, while the patient's learned response was correct in one circumstance, her lack of explicit memory had robbed her of the ability to adapt the response for other circumstances. When there was no danger, she still jerked her hand back. A normal person would have laughed and said, "Now, Doc, you don't have a pin in your hand this time, right?" But she'd lost the flexibility that makes *Homo sapiens* unique in the animal world.

Voluntary actions on the right side of the body begin in the motor cortex in the left hemisphere of the brain and go through the pyramidal tract, a great bunch of axons that issues from it. If you have a stroke that destroys the motor cortex, you'll be paralyzed on the right side of your body. Everything will stop working, including your facial muscles. If someone tells you to smile, you'll produce a grotesque, lopsided grimace. But if you hear something funny that causes you to laugh involuntarily, you'll produce a normal, symmetrical smile. The reason is that emotional reactions are controlled through the anterior cingulate, the medial temporal lobe, and the basal ganglia, which were not destroyed by the hypothetical stroke. That effect can be reproduced in patients with brain damage. There are at least two separate brain systems that can generate behavior. The way they work, the way you capture experiences and turn them into learning (memories), can influence your ability to survive.

M. Ephimia Morphew, a psychologist and founder of the Society for Human Performance in Extreme Environments, told me of a series of accidents she'd been studying in which scuba divers were found dead with air in their tanks and perfectly functional regulators. "Only they had pulled the regulators out of their mouths and drowned. It took a long time for researchers to figure out what was going on." It appears that certain people suffer an intense feeling of suffocation when their mouths are covered. That led to an overpowering impulse to uncover the mouth and nose.

The victims had followed an emotional response that was in general a good one for the organism, to get air. But it was the wrong response under the special, non-natural, circumstances of scuba diving. It's possible that the impulse, the feeling of suffocation, was formed as an implicit memory by some previous experience that was not available to conscious (explicit) memory. And the divers had no way of knowing that the one thing that would keep them alive, covering the nose and mouth, was the one thing the organism would not tolerate. At the critical moment of decision, reason was not enough to overcome emotion. For no one would say that those divers believed they could breathe under water without a regulator.

Morphew and the other researchers wanted to know what the divers were thinking when they removed their regulators and tried to breath without them. The answer is: you don't need to think. That's what emotions and implicit memories are all about. By tradition, reason is regarded as the highest function. People are named after it: *Homo sapiens* (from the Latin *sapere*, to taste, as in "to taste the world"). But from the point of view of an organism in desperate trouble, an organism that evolved by relying on emotions as the first line of defense, cognition is irrelevant and gets set aside. It's slow and clunky. As Remarque said, there's no time for it.

Most of the mystifying accidents that happen in the course of risky recreation, the seemingly illogical decisions, actions, and outcomes, can be explained by the same interplay of emotions and cognition that shapes all human behavior. What the scuba divers did made perfect sense from the point of view of the organism's survival: The impulse to get air is automatic and can be overpoweringly strong. Those who can control that impulse to survive, live. Those who can't, die. And that's the simplest way to explain survival, whether the venue is night carrier landings or being lost in the jungle.

When I was teaching my daughter Amelia how to snowboard, she caught her edge exactly twice and never did it again. When you catch the downhill edge of your snowboard, it slams you to the ground with such force you feel as if somebody just dropped a safe on you. It hurts everywhere. After the first time you do it, you have to think consciously about tensing the muscles in your legs to keep your edge up as you ride. Then you get tired, lazy, or distracted, you catch your edge again, and it practically knocks you out. Generally, after that second experience, you have developed a deeply ingrained emotional bookmark. Then, whenever you start to relax those muscles, you get a really bad feeling, like somebody's going to drop a safe on you, and those muscles tighten right up. You never have to think about it again.

The elegant and seamless assistance of those bookmarked feelings is essential in using the more linear tools of logic and reason. But in certain circumstances, usually ones where we are exposed to unfamiliar or extreme hazards, it can also be a trap.

THREE

A MAP OF
THE WORLD

WHEN I WAS a teenager and my father could see how hard I was trying to be cool, he took me aside and gave me a little talk about the difference between being cool, which he clearly was (though he didn't say that), and acting cool, which I clearly was. I was already doing risky things, and he remarked that to die a needless death of your own making was not cool, man, not cool.

One summer, some years later, I was going out on night assault maneuvers with the Army's 82nd Airborne Division at Fort Bragg in a live fire exercise. I was working with Army Rangers, who were some tough mothers. We spent a few days hopping in and out of helicopters, shooting off rockets, and blowing up tanks. Oh, it was great fun.

But I was most interested in hearing from the Rangers about their training. It was hard core. The training lasted eight weeks, starting at Fort Benning in Georgia and winding up sometimes as far away as Utah. It was the most intense, demanding, and exhausting ritual the U.S. Army had.

We'd just come on a night drop into sand dunes. They didn't let me jump, so I was on the ground when the planes came over. I could hear the distant chattering of automatic weapons fire somewhere in the forest and then the faint droning hammer of the C-130 engines. The moon had not yet risen, but the Big Dipper was up in the western sky, and a big orange planet burst halfway up beyond the southern treeline as I walked out onto what seemed like desert but had been, 10 million years before, ocean floor. The darkness was so complete that I could see nothing before me as I struggled through the sand. Then one of the stars in the southern sky resolved itself into an artificial light, and I walked toward it, knowing that it was a turn-in point for parachutes and that people would be there.

As I walked toward it, I saw a planet suddenly rise half out of the eastern sky, and I watched it for a moment before I realized that I was looking at a flare. A moment later automatic weapons ripped away, and then the heavy *whoosh-thud* of a howitzer peeled back the pretense of solitude for a moment before night closed once more around me. There is nothing quite like the sound of a howitzer—an enormous, galvanized steel door being slammed, jamming the air up into the valleys of these hills. Like a surf, the waves of air come back when they've spent all their energy out there.

Then the planes came, and I saw their green and red navigation lights moving toward us along a line parallel with the drop zone. I ran to get under them as they came on and on. I reached the middle of the DZ just as they drew overhead. Everything seemed to grow silent as the stress of the moment narrowed my perceptions. The sky was light compared with the land, and against that shimmering, cold, feather-gray scrim I saw the dark leviathan shapes of the ships crossing to the south. Without warning, a blossoming profusion of jellyfish sprayed out across the sky. Silently and swiftly they grew from black points in the sky to the swelling, round, living atoms of darkness, filling in the spaces between the stars.

Now the planes were gone, and truly there was no sound at all except my heart hammering. As I stumbled on the ocean floor, watching scores of the creatures come down all around me, I knew that one must surely drift down on top of me and engulf me in the trembling petals of its mushroom flesh. I could see, as they descended in the fluid of the air, that men were dangling from them. Within 100 feet of the ground, each man pulled the release that dropped his rucksack to dangle on a 15-foot lanyard, and all around me now I heard the snap-clatter-swishing of the packs as they hit and the men prepared to land.

Just beside me, the first man landed hard, and I heard his "Oof!" and saw the gray jellyfish above him balloon and invert, dumping its bubble of air, then drift and fold and lie down quietly on the sand. "Oh, God, I've got to piss!" the man groaned, and I heard the clanking of his gear as he tried to free himself of hasps and clips and webbing.

His face came up and I could see him just well enough in the dim light to recognize him. It was a guy they were calling Buddy. When we were gearing up for the drop, zipping speed loaders into M-16 clips, spacing them with red phosphorous tracer rounds, and unpacking mortar rounds to stuff into butt packs, I'd seen Buddy down on his knees, cradling a Claymore mine. A Claymore is a modern version of the stand of grapeshot, a fiberglass shell filled with plastic explosive mixed with seven hundred steel ball bearings. It's rectangular, about 6 by 10 inches, 1½ inches thick, dull silver colored, curved from side to side, with pointed metal legs that fold down from underneath so that it can be stuck into the ground like a miniature drive-in movie screen. When it's fired, that load of grapeshot sprays out, supersonic and molten, and anyone within a 325-foot swatch is reduced, as the Rangers liked to say, to Hamburger Helper.

I'd been playing with the switch on a mortar round, flipping it from proximity to impact, when I looked over and saw Buddy with his face against the Claymore, a cigarette in his mouth, smelling it. I went over to see what madness motivated him to do that, and he looked up, smiled at me dreamily, and said, "Mmm, I love the smell of a Claymore." I got down and cradled it to my face and smelled it, too. It smelled like cherries.

Now on the DZ, as all the men landed around us—first the ruck-sack's crunch, then the man, hitting and rolling as best he could, encumbered as he was with live rounds, rockets, high explosives, Claymores, grenades, flares, and then the great weight of the para-chute itself, the sea creature that carried him here, dying in the dead air on the ancient shore, I heard one man say as he drifted down toward me, "Look out, son. I'm gonna 'splode on impack."

Then I was humping through the forest in the middle of the night with a 40-pound pack, talking with Colonel Robert Lossius, who'd gone through it. He was a Ranger. This night, for him, was just a picnic. He told me that after a week or two in Ranger train-ing, he'd begun to hallucinate, an experience often reported by survivors. The Colonel's ration had been two MREs (meals ready to eat) every three days, which is to say he was slowly starving. He lost 30 pounds in fifty-eight days. There was sleep deprivation and an escalating series of trials—forced marches, free-climbing sheer cliffs without ropes, and being thrown into the snake- and alliga-tor-infested swamps in Florida to find his way out on his own. Fear of death, fear of height, fear of dark, fear of drowning, all com-bined to make him descend deeper and deeper into himself and find what was there, to change it, to build it. Some say to put it to sleep. Others say to wake it up. I left Fort Bragg thinking that of all the people I'd known, certainly a Ranger would be the most likely to survive the hazards of nature. But the training is only as good as the environment.

ON SEPTEMBER 6, 1997, Captain James Gabba, an Army Ranger, was taking a guided commercial rafting trip down the upper Gauley River in West Virginia when the raft hit a rock. Gabba, thirty-six, was thrown from the raft, and his guide, trying to save him, fell in too. The guide tried to rescue Gabba, but Captain Gabba "just laughed and pushed him away." Gabba floated calmly downstream, and if he was anything like the Rangers I'd known, he must have felt that he was in no real danger because of all the training he'd had under much worse conditions. He must have felt good, too, masterful, confident. He'd eaten, he'd slept—hell, he could do this until the wheels came off. Then he arrived at a place where a big rock blocked the middle of the current. Gabba was sucked under, pinned, and drowned. The official report said, "The guest clearly did not take the situation seriously." But that's not true. He took it very seriously.

It's easy to see hubris in Gabba's behavior, but it's more subtle than that. Everyone carries around a necessary measure of his environment and of the self. From conception onward, the organism defines what is self and what is not. The immune system examines materials from the environment to assess whether they are a threat or harmless. Cells within that system hold up proteins in an almost ritualistic way so that T-cells can read them and see if they are self or foreign. If a T-cell recognizes the protein as self, it commits suicide. If the protein is unfamiliar, the T-cell gives the B-cell permission to create an antibody, which helps mobilize an attack to destroy the invading protein.

In that and other ways, the immune system continuously rearranges the organism's relationship to its environment. That's called adaptation. A lifetime of experience builds the system, but a subtle change in the environment can mean that the system no longer has the correct response. It's suddenly out of adjustment. For example, when Europeans brought unfamiliar diseases west with them in the 1500s, the previously healthy and thriving Native Americans were rapidly wiped out. But some creatures are amazingly adaptable. At the beginning of summer, I used to have hundreds of crows circling my house at dawn, barking noisily. Then one morning they were all gone. They'd been threatened by the West Nile virus. They went away for months. Now they're back. Crows are survivors.

The emotions are another mechanism for defining self (actually creating the self) during the process of protecting what is within from what is without, both by avoiding or fighting what is bad and by seeking out what is good. As Joseph LeDoux put it, "People don't come preassembled, but are glued together by life." Like the immune system, the emotional system evolves continuously, taking experiences and situations and attaching emotional value to them in subtle gradations of risk and reward.

Children begin learning even before birth, and near the end of pregnancy their brains may be forming as many as 250,000 new nerve cells a minute. (Scientists estimate that the mature brain has 100 billion neurons and trillions of connections.) Once infants begin moving about in the world, they engage in a process of trial and error by which they find out how much risk they can take to reap a given amount of reward. Every experience adds to the body of knowledge and shapes future behavior. Children constantly test and sample their environment and themselves, taking risks that give big rewards without too much exposure. It's a delicate, often beautiful, balancing act. I watched my newborn son, Jonas, learn to cry on purpose. In animal terms, you can view crying as a risk, because it attracts attention, and there's no way for the infant to know if that attention is going to be good or bad. At first, he'd only cry when something bothered him—discomfort, pain, hunger— and you could tell by the sound of the cry that it was genuine (and loud). Each time he cried, however, his mother held him and most times she fed him, too, or changed his diaper. Soon Jonas learned that if he wanted some sort of attention, he could cry, and it was a different cry with none of the urgent, shrieking qualities it had before. It was more a whimper, and he used it effectively to get what he needed or wanted.

Moderate stress enhances learning. When two neurons fire together, they become wired together. When a strong and weak neuron—call them Al and Betty—stimulate a third neuron—call it Charlie—at the same time, the weak one, Betty, gains the ability to stimulate Charlie to fire. That's why the ringing of a bell could cause Pavlov's dog to salivate even when there was no food present. Scientists, with their ever playful juggling of three or four languages at once, call that long-term potentiation (LTP). So risk is an integral part of life and learning. A baby who doesn't walk, for example, will never risk falling. But in exchange for taking that risk, he gains the much greater survival advantage of being bipedal and having his hands free. That's another reason play becomes important to most people. That's why we go out there and do the things we do in the wilderness. Amelia could have refused to snowboard after her first painful fall. But in exchange for the risk, she gets to have the emotional rush of zinging down a mountain and catching some phat air with the old man. She also gets to wear cool snowboarding clothes to attract boys, so perhaps there is survival value in it, too, at least for the species.

The knowledge gained from that risk-reward loop does not involve reasoning. It comes to the child coded in feelings, which represent emotional experiences in a particular environment. If the environment changes, if it has unfamiliar or subtly different hazards, those adaptations may turn out to be inappropriate.

Logic simply takes too long, often impossibly long, and in a child logic is not well developed enough at any rate. Instead, he rapidly and unconsciously pages through his atlas of emotional bookmarks (probably an instance of LTP). Numerous neural networks, associations that connect the situation he's in with similar situations or experiences from the past, flicker with electrochemical energy, illuminating memories and feelings of circumstances and actions that led to good and bad outcomes, projecting forward to future paths of action and feeling. Jonas is forming a dynamic map of himself, his world, and his experiences in it, and is sending projections through time in the form of images of the future. In a sense, he can see down his own path by the light of that circuitry. The emotional map of the world, with him in it, is doing its work all the time, essential as a heartbeat to his survival.

His work had only just begun. Years later, he might be sitting quietly at home, and that system will be doing another kind of work, perhaps helping him to decide whether to read *War and Peace* or eat ice cream. If he ever falls into a fast-moving river, that system will be doing a much more urgent sort of work. And if he's had the right experiences, it will instantly direct correct action.

IF YOU could see the brain working, if it gave off light as it worked, then at a decision point, different areas would begin glowing all over like the lights of cities going on at dusk as seen from space. The patterns contained in those networks, formed from unique experiences of life, inform decisions at a rate of speed that can never be achieved by logic. And all of this takes place in the shadowland just beyond conscious thought.

Most people who fall into swift water would welcome rescue. But Captain Gabba, the Army Ranger, may have set himself up for disaster through extreme experiences in extreme environments. As a Ranger, Gabba had had experiences that seemed more hostile than a guided river tour. He'd not only taken care of himself, but if the stories told me by other Rangers are any indication, he came away feeling more alive than he'd ever felt. Worse, in Ranger culture, having to be rescued is ignominious. It is associated with a bad outcome: shame, failure. In Ranger training, if you have to be rescued, you are out of the program. The emotional bookmarks that Gabba developed had labeled rescue as bad and self-sufficiency and even pain as good, no matter how threatening the environment. His training and experience taught him that it was better to die for his country than to fail. Death before dishonor. Rangers lead the way, they don't follow. The training worked.

The Canadian snowmobilers suffered a similar consequence stemming from emotional bookmarks. The first snowmobiler who went up the hill had the abstract concept of an avalanche idling around in the front part of the brain, looking for something physical (i.e., emotional) to attach itself to. Unfortunately, it found nothing in the atlas of experience. His emotional map of the world contained no feelings about avalanches because his body had had no experience of them.

On the other hand, he had clear perceptions to help find an emotional bookmark: the mountain and environs, the noisy snowmobile. For example, he may have smelled the pine forest and it was the same smell he'd noticed the last time he tried hammerheading, and when he did, he got this boss feeling. He had the bodily memory of pleasure from previous runs. One was an abstraction, the other a physical certainty, clearly illuminated within his emotional map and signaling deeply instinctive behaviors ("If you don't hunt, you don't eat," for example).

Certainly, there were other factors. The masterful feeling of being the rescuer and the speed of riding ahead to that beautiful spot in the woods, not to mention, perhaps, fatigue, dehydration, an earlier bout with anxiety about their lost friends, the relief of finding them (the "Whew Factor," in which you let down your guard once you feel safe). All of those influences, too, must have conspired to derail the efforts of reason to constrain action.

There was another more fundamental difficulty that the snowmobilers faced. Our sense of a mountain, the earth, is a sense of something solid, and our experience confirms that. Nothing in our learning tells us that a mountain is going to come apart before our eyes. It makes no sense. It hasn't happened, therefore it cannot happen. The mountain certainly didn't look fragile. The snowmobilers literally couldn't believe it. We think we believe what we know, but we only truly believe what we feel.

ONLY IN recent years has neuroscience begun to understand the detailed physiology of emotional states such as fear. The neocortex is responsible for your IQ, your conscious decisions, your analytical abilities. But the amygdala stands as a sort of watchdog for the organism. Amelia, who is the younger of my two daughters, has a chocolate Lab, Lucy. Lucy sometimes reminds me of the amygdala: When anyone comes to the door, she barks before I even hear it.

Perceptions from the world around us (sight, for example) reach the thalamus first. In the case of vision, axons from the retina go to the visual thalamus (there are two, one in each side of the brain, receiving information from each side of the body). From there, the signals travel by way of axons from the visual thalamus to the middle layer of the neocortex and from there are sent out to the other five layers for processing. What emerges is a perception of sight. But before all that can be completed, a rough form of the same sensory information reaches the amygdala by a faster pathway. The amygdala screens that information for signs of danger. Like Lucy, the amygdala isn't very bright, but if it detects a hazard, or anything remotely resembling one, before you're even conscious of the stimulus, it initiates a series of emergency reactions. The approach is: Better safe than sorry. (Unlike Lucy, the amygdala also is capable of ignoring a lot of information as irrelevant.) It is a primitive but effective survival system that causes the rabbit that visits our backyard every morning to freeze and then run when she sees Amelia let Lucy out. Like Lucy, the amygdala is wrong a lot of the time: There is no danger. But in the long course of evolution, it has been a successful strategy.

So information from the senses takes a neural route that splits, one part reaching the amygdala first, the other arriving at the neocortex milliseconds later. Rational (or conscious) thought always lags behind the emotional reaction. Anyone can demonstrate this at home: Everyone has been startled by someone. It's a powerful response, marked by the familiar rocket rush of adrenaline (actually catecholamines), increased heart rate, flushing, and panting. Then, as soon as you realize the person is someone you know, the response deescalates. But it takes a while to metabolize all those chemicals. It's a powerful emergency reaction and completely illogical, because you know the person and are not in any danger.

But you can't think of that logically before reacting because the visual signals reach the amygdala first. It's a big shadowy form: It could be a spouse, it could be a bear—you don't know. Only later (in milliseconds) does the visual cortex piece together an accurate picture that lets you in on who it is. Only later can you reason: No bears in this house.

While the pathways from the amygdala to the neocortex are stronger and faster than the ones going the other way, some ability may remain for the neocortex to do the following: First, to recognize that there is an emotional response underway. Second, to read reality and perceive circumstances correctly. Third, to override or modulate the automatic reaction if it is an inappropriate one; and fourth, to select a correct course of action.

Since emotions evolved to elicit behaviors in a split second, clearly, that is a tall order, and some people are much better at it than others. In addition, there is wide variation in individual reactions. Some people startle easily. Others tend not to react at all. Some people function better under stress, such as professional golfers, fighter pilots, elite mountain climbers, motorcycle racers, and brain surgeons. And some emotional responses are more easily controlled than others.

Elite performers, as they're sometimes called, seek out the extreme situations that make them perform well and feel more alive. At the other end of the scale are people who don't want any excitement at all. It takes all kinds. But it's easy to demonstrate that many people (estimates run as high as 90 percent), when put under stress, are unable to think clearly or solve simple problems. They get rattled. They panic. They freeze. Muddled thinking is common in outdoor recreation when people get lost or injured or are otherwise threatened with harm.

But even elite performers are not immune to the effects of stress. Greg Norman completely blew the 1996 Masters golf tournament after developing a crushing six-stroke lead over Nick Faldo. He missed a three-and-a-half-foot putt, knocked a ball into the water—

twice—hooked balls, missed another putt . . . it was brutal. At the end, Norman and Faldo could only hold each other and weep.

When you learn something complex, such as flying, snowboarding, or playing tennis or golf, at first you must think through each move. That is called explicit learning, and it's stored in explicit memory, the kind you can talk about, the kind that allows you to remember a recipe for lasagna. But as you gain more experience, you begin to do the task less consciously. You develop flow, touch, timing—a feel for it. It becomes second nature, a thing of beauty. That's known as implicit learning. The two neurological systems of explicit and implicit learning are quite separate. Implicit memories are unconscious. Implicit learning is like a natural smile: It comes by way of a different neural pathway from the one that carries explicit memory. LeDoux reports that his mother, who has Alzheimer's disease, cannot remember ordinary events but can still play the accordion, because although her hippocampus is likely damaged by the disease, the memory of how to play accordion comes from an as yet undamaged part of the brain. Implicit memories are not stored in or necessarily even available to the analytical, reasoning part of the brain.

In a normal person and under just the right conditions of stress—perhaps you're tired, perhaps you're getting a cold, perhaps you're going through a divorce—that implicit system can break down. Then you're left with the explicit system, thinking through each motion like a rank beginner. Malcolm Gladwell, writing in the *New Yorker*, put it succinctly: "Choking is about thinking too much. Panic is about thinking too little."

I flew aerobatics for a number of years and competed with the International Aerobatics Club. One day my instructor and good friend Randy Gagne crashed with a student on board. They went into the ground going perhaps 250 miles an hour. Until then, I thought I had some special dispensation to fly upside down with impunity. After that accident I realized I was not an elite performer. Most people aren't. They're just out there to have a good time. And when things go wrong, they have no idea what's happening to them. It may seem as if we live and die by dumb luck, but it is much more subtle and complex than that.

Everything was stacked against those snowmobilers, including the way their brains were organized and shaped by continuous adaptation to their environment. Those were not stupid people, nor were they ignorant, nor even necessarily reckless.

At the decision point, with all those neuronal networks lighting up, fixing their focus, and arming the mechanism of physical action, the one clear and certain thing was that riding up the hill would produce a good feeling and one that, to the organism, seemed necessary, for it arose from emotion aimed at ensuring survival. Since they were already all pumped up, the concept of idleness as a haven of safety couldn't compete with the feeling of motion as survival. To the organism, the decision was clear. The emotional bookmark, that "beacon of incentive," burned brightly, and the decision was made in an instant, outside the realm of conscious thought.

As Captain Gabba laughed and pushed his rescuer away, he too must have had a good feeling about what he was doing, just before he was pinned and drowned.

FOUR

A GORILLA
IN OUR MIDST

THE ILLINOIS RIVER in southwestern Oregon has thirty-five miles of class III to IV rapids with a class V, moss-covered gorge in the middle. That section is known as the Green Wall. Gary Hough, a minister on holiday, knew that with his level of skill, he could run the Illinois at flows between 900 and 3,000 cubic feet a second. He also knew that half an inch of rain would be sufficient to bring the flow above 4,000 cubic feet a second, at which volume the Illinois would be too rough for all but the most prodigious paddlers, who preferred to run it at 2,500.

Conditions looked right, if just barely, that Saturday morning when his group began the trip. The flow was 2,000 cubic feet a second. Although a storm was predicted for late Sunday, Hough estimated that they'd have enough time for the three parties attempting the run to get through. He was willing to accept a certain level of risk for the reward of the trip. He had a clear idea of the dynamic forces involved in the environment he was about to enter. He had a reasonable conception of his ability to deal with it, and he knew what changes in the system would overpower his skills.

Shortly after Hough's party set out on March 22, 1998, it began to rain. He did the right thing: He ordered the craft pulled out and had his group make camp. It wasn't much like the kind of fun they had planned. But then again, the water was rising so fast they had to move their craft several times to keep them from washing away. Conditions in the real world, the objective hazard level, had exceeded the risk Hough was willing to accept. The environment had changed, and he adapted. Using his reason to manage emotion and emotion to inform reason, he survived.

Between March 21 and 23, 3 inches of rain fell, melting snow in the Siskiyou Mountains, adding even more water to the flow. The river rose 15 feet and the flow eventually would reach 20,000 cubic feet a second. According to Charlie Walbridge's *River Safety Report*, "this surge caused havoc among weekend river runners."

"Anybody who's paid attention to a flooding river will know it," Hough later said of the sound that woke him the next morning. "There's the roar of the full-throated river, but on top of that, as if it's a layer you could pick up and remove, there's the hiss. That hiss basically says, 'Keep your distance.'" That's a pretty subtle cue, I suppose. But to the survivor's mind, all cues are important. They carry information. So a survivor expects the world to keep changing and keeps his senses always tuned to: *What's up?* The survivor is continuously adapting. That hiss was enough for Hough.

For those who were not so attuned, there were other cues: Dawn revealed the python that had come in the night to swallow the Illinois whole. "When the river has been coming up a foot an hour all night, when it's gone from clear to chocolate milk, when there are no more eddies and there are 18-inch diameter trees going down at 15 miles an hour, it's just not a tough decision," Hough said. But as John F. Kennedy once remarked, "There's always some son of a bitch who doesn't get the word."

Hough's party, then, watched from the refuge of camp as a party of five on a shredder (a type of raft) and three kayaks came shooting past at around 10 A.M. About half an hour later, they were followed by a second party of five, including at least one kayak. The first group, a man and a woman in the shredder, were flipped into a 12-foot standing wave after portaging the Green Wall. The second group managed to portage the Green Wall, but two paddlers were eaten near the Little Green Wall, another hellacious section. They were still operating on a model of the old environment. The results were fatal: One member of each party drowned, and the others lost all their equipment and had to be rescued. One survivor, the woman who had been in the shredder, was carried five miles downstream before she was able to get out.

Walbridge wrote, "Coast Guard rescue helicopters picked up ten people, six of whom had flipped boats and become stranded in a sheer-walled section of the canyon." The first man killed, Jeff Alexander, thirty-seven, was an experienced river guide. The second, Wilbur Byars, sixty-two, was also a well-known river guide who had many years of experience. It is unlikely that professional guides would intentionally take such a risky run. In fact, they spent some time at the put-in discussing the risks, and everyone agreed that the river could be run. They had a lot of experience at running the river, and it had always been okay. Big water, said their emotional systems, equals big fun.

AS COMPLEX as the brain is, the world is more so. The brain cannot process and organize all the data that arrive. It cannot come up with a reasonable course of action if everything is given equal weight and perceived at equal intensity. That is the difficulty with logic: It's step-by-step, linear. The world is not.

Perceptions come at you like the 6 million hits you get when you do an Internet search. Without a powerful search engine, you're paralyzed. One search engine involves emotional bookmarks, in which feelings help direct logic and reason to a place where they can do useful work. A second strategy the brain uses for handling complicated problems is to create mental models, stripped-down schematics of the world. A mental model may tell you the rules by which an environment behaves or the color and shape of a familiar object.

Suppose you're searching the house for your copy of *Moby-Dick*, and you remember it being a red paperback book but you don't know where you left it. When you search, you don't examine every item in the house to see if it's *Moby-Dick*. That would be logical, a strict use of the faculty of reason. But it would also be tedious and would take too long. That's how a computer would do it. The fact that you have a mental model of the red paperback copy of *Moby-Dick* allows you to screen out nearly everything you see until, at last, a red book blossoms in your field of vision. But if you're wrong and it's a blue hardback edition of *Moby-Dick*, chances are that you won't find it even if the title comes into view.

Everyone is familiar with finding something "right under my nose." A faulty mental model is part of the explanation. It's the reason that you can get off an elevator on the wrong floor. It's the reason that many card tricks and magic acts work: You see what you expect to see. You see what makes sense, and what makes sense is what matches the mental model. If you do succeed in finding your copy of *Moby-Dick*, your pupils will dilate at the moment you recognize what you're looking for, as they do when you reach the solution to a mathematical problem or see something you like.

My father used to perform a card trick for me and my brothers. Anyone can do it. He'd shuffle a deck and ask us to pick a card, not show it to him, and replace it anywhere in the deck. Then he'd let us shuffle the deck so that it would seem impossible for him to find the card. But when he'd show us the cards, one by one, he'd watch our eyes. When we saw the right card, our pupils would dilate, and he'd know which card we'd picked. The involuntary physical response is proof that there is an emotional component to the process of matching the model with the world.

In more subtle tricks, the magician creates a mental model for you, a short-term memory of the world. Every model of the world comes with its own underlying assumptions based on experience, memories, secondary emotions, and emotional bookmarks, all of which influence what we expect to happen and what we plan to do about it. Then the magician switches realities, and while you stick with the model that is closest to your familiar reality, he reveals the new reality. The disconnect is what's surprising. You believe the magician does the trick, but in fact you do it yourself. Magic is astonishing only to the extent that you ignore how the brain works. It's a bit of a magic trick, then, to put a boat in a river when 18-inch trees are shooting down it at 15 miles an hour.

One of the reasons magic tricks work can be explained through a system called working memory. It is a general purpose workspace, and most of us experience it as attention or conscious thought. In addition, there are specialized systems for verbal and nonverbal information, and they have a type of short-term memory that allows perceptions to be compared with one another over the span of a few seconds. The general purpose area can take in information from the specialized systems (sight, smell, sound, and so on) and can integrate and process that information through what LeDoux calls "an executive function." That area of the brain, located mostly in the frontal lobes, is responsible for making decisions and voluntary movements, as well as directing what sensory input we're paying attention to. It's why we can still carry on a conversation in a room where many people are talking and music is playing. It's why we can choose between getting up and putting on a sweater or turning the thermostat up.

As LeDoux and others have explained, working memory can hold only a few things at once, perhaps half a dozen or so, and when something new commands attention, those things are forgotten. Working memory can also retrieve information from long-term memory. The fact that you can read this long sentence is the result of your working memory's ability to hold the beginning, middle, and end all at once and to retrieve definitions and associations from long-term memory and use them to make sense of the words. It is also the result of the fact that you have created mental models of the words. You don't read each letter to decode the word, as a child who is learning to read must. But if you come across words that are too similar, such as psychology and physiology, you may have to pause.

The fact that new information, especially emotionally charged information, forces things out of working memory means that we can't pay active attention to too many things at once. Unless something is successfully transferred from working memory into long-term memory, it is lost to conscious awareness. We all have this experience when we try to memorize something that has no emotional content, such as an address or driving directions. In most people, the executive function can do one task at a time, and attempting to perform simultaneous tasks that involve a conflict begins to break it down. For example, if you flash the word "blue" printed in green ink on a screen for a second and then ask someone to say the word or the color, he'll have to stop and think before he answers.

The limited nature of working memory (attention) and the executive function, along with the shorthand work of mental models, can cause surprising lapses in the way we process the world and make conscious or unconscious decisions. That is why even experts can miss things that are right under their noses. In May 1989 Lynn Hill, one of the most famous rock climbers in the world, was in Buoux in southern France preparing to climb a one-pitch bolted route. Bolted routes are where beginners start, because they are the safest of climbing environments. Hill and her husband were simply climbing it as a warm-up. She became distracted as she was tying the rope to her harness and didn't finish her knot. The task of tying the knot had been in working memory when a conversation with a Japanese woman nearby and the act of putting on (and tying) her shoes bumped it out. (With conflicting demands on working memory, she may have unconsciously substituted tying her shoes for tying the climbing rope.)

As Hill walked back to begin her climb, there was a slight residue of the memory left, but it had not been completely formed. "The thought occurred to me that there was something I needed to do before climbing," she wrote in *Climbing Free*. The fragments of a half-formed memory had left a gut feeling, an emotional bookmark, and she considered taking off her jacket, which would have revealed the half-tied knot, but "I dismissed this thought."

She climbed to the top unaware of the trap she'd set for herself. She had created a mental model that had worked in the past. She was now depending on it to match reality. She asked to be lowered, put her weight on the rope, and the half-tied knot came loose. She fell 72 feet. A tree barely saved her.

CHARLES PERROW is a sociologist known for studying industrial accidents, such as those that occur with nuclear power plants, airlines, and shipping. In *Normal Accidents,* he wrote that "We construct an expected world because we can't handle the complexity of the present one, and then process the information that fits the expected world, and find reasons to exclude the information that might contradict it. Unexpected or unlikely interactions are ignored when we make our construction." The snowmelt in the Siskiyou Mountains, the unseasonal rains, were unexpected and unlikely interactions in a system involving water, air, gravity, and terrain faced by boaters on the Illinois River. The possibility that the river could rise 15 feet to reach a flow rate of 20,000 cubic feet per second was outside the normal experience of the river's guides, contradicting their model. It may seem difficult to believe that they could have missed the trees zooming down the river and the noise it was making, but it might be easier than it seems. In fact, the two people in the shredder were caught for a time in a violent eddy, along with a floating tree.

Mental models can be surprisingly strong and the abilities of working memory surprisingly fragile. A psychologist who studies how people behave when they're lost told me, "I saw a man I was hiking with smash his compass with a rock because he thought it was broken. He didn't believe we were heading in the right direction."

Perceptions do not incorporate the world in a literal way. If you try to imagine the face of your mother, there is no photograph of your mother waiting in your brain. Rather, fragments of the image, including its emotional content, are stored in a dormant form scattered across numerous neural networks. When you call forth your mother's face, they all light up simultaneously in what Antonio Damasio calls a trick of timing to create an illusion that you're seeing your mother's face, or something like it. The way we store what we know about the world is much the same. Mental models are distributed across the brain.

Jacques Barzun wrote about a man named William Huskisson, a member of Parliament for Liverpool, who was attending the opening of the Liverpool & Manchester Railway on September 15, 1830. "At the inauguration . . . [he] was run over by the *Rocket*: no one as yet knew how to gauge speed and distance, when and where to step. Man's reflexes failed him in the midst of powers stronger than the horse and the ox."

Huskisson's mental model of his world had served him well. He knew how to cross a street crowded with vehicles. They were powered by horses, and he didn't need to calculate their speed. It was second nature to know how much lead time to allow himself to get across. Then his environment changed suddenly. He had never seen anything move as fast as a locomotive. Huskisson looked, saw where the train was, and unconsciously consulted his mental model to rate risk and reward. He then estimated (again, outside of consciousness) that he could cross the tracks in time. But his model was rendered useless by a change in the environment brought about by an invention.

I don't know what Huskisson was thinking, but since the *Rocket* was the first commercial locomotive, it's possible that he had not seen a train before. If, as a business leader, he had seen a train, it was new information, capable of altering his mental model, yet surely not as firmly embedded as the old, and not properly bookmarked for quick reference. His was a model in transition of a world in transition. It's also likely that excitement was high that day. A Tory Prime Minister and the Duke of Wellington were in attendance, along with Prince Esterhazy, who was very nearly killed too. Railroads were about to be big business, and as a controversial free trade advocate, Huskisson must have had a lot on his mind. The pressure was on, the crowd excitable. (After the accident the crowd stampeded the tracks and tried to attack the infamous Duke of Wellington, who refused to board a train again for more than a dozen years.) Huskisson happened to see the Duke of Wellington on another train full of dignitaries, and although estranged from him, he decided to cross the tracks and greet him. So we can imagine that in the excitement and stress of the moment, Huskisson would have been having difficulty holding the important things about his immediate environment in working memory and integrating the conflicting inputs. He was screening out significant information at the decision point, reverting to old familiar models of his world. Older bookmarks may have set the trap. They are, after all, slaved to emotion, which triggers action. His feet moved. All it took was a step or two. Then he was in front of the engine, perhaps recognizing his error only a second too late.

WHEN AMELIA and I were first learning to snowboard, we became disoriented in a blizzard at a ski resort in Washington State. We had little experience and less skill, and the other resorts we'd been to, like most, had the lodge at the bottom of the lifts. We saw a long line of people trying to get on a lift. Of course, I knew that the lodge for this ski resort was at the top of the hill, not the bottom. We'd been there for breakfast. But with the stress of exhaustion, dehydration, emerging hypothermia, and the fear I felt for Amelia's safety, I ignored those cues. If I thought about it at all, I'm sure that I couldn't figure out why they were going up again in such a whiteout.

Stress doesn't take long to confuse you. We'd only been out a couple of hours. I kept pushing Amelia downhill, away from safety, off into the wilderness. I was able to blank out the evidence of my senses: I could see people going up the lift. I relied on the familiar model of a ski resort, spurred on by the gut feeling that danger was up, safety down. Fortunately, Amelia, who was only twelve years old at the time, was more attuned to her environment, and even in the whiteout, she kept her sense of direction. "Pop," she said. "The lodge is up, not down."

Oh, no. *What was I thinking?*

Half frozen, we climbed back to the lift, rode up to the lodge, and took the gondola down to our car on the other side. I'm grateful that she knew how to walk her own walk. It would have been a long night out.

MENTAL MODELS, emotional bookmarks, and the ability to keep the right things in working memory played a powerful role in determining who lived and who died in the collapse of the World Trade Center towers. When the first airplane hit the north tower, it cut off all the stairwells and elevators, trapping anyone above the ninety-first floor. When the second airplane hit the south tower, a stairwell on the northwest side of the building remained open, despite the extensive destruction throughout the area from the seventy-eighth to the eighty-fourth floors. While eighteen people got out by using those stairs, at least two hundred went the other way, heading for the roof. A fire warden named Brian Clark stood in the stairwell with his flashlight, asking people as they entered, "Up or down?" And while a few went down into the rising smoke (and escaped), most went up. With only fourteen minutes left before the collapse, hundreds of people were standing at the wrong end of the stairwell facing the locked door of the roof.

Some had followed Roko Camaj, a window washer who had a key to the roof door. Others relied on a powerfully emotional past experience. When the World Trade Center was bombed in 1993, people had been short-hauled off the roof by helicopter, creating a model of safety and escape that led many up, presumably toward air, not down into smoke. Camaj found that his key wouldn't unlock the door. They were trapped. And even if they had all started down at that point, there was not enough time left to reach the street before the collapse. Their previous experience or the shared experience of others betrayed them, and they all died.

While it's understandable that people could be confused or make wrong decisions under the extremely unusual circumstances in the World Trade Center on that day, it's more baffling that I would fail to grasp something as obvious as the location of a ski lodge or that river guides could miss huge trees shooting down a river. The guides were, after all, the ones who died. The clients lived. None of us was on a suicide mission. No, they (and I) had profound lapses in our ability to use working memory properly, to process new information from the world, and to integrate it with long-term memory. And there's plenty of evidence that such lapses are perfectly normal and happen all the time.

Arien Mack, a psychologist at the New School for Social Research in New York City, writes that "Most people have the impression that they simply see what is there and do so merely by opening their eyes and looking." A group of psychologists at Harvard performed a rather disturbing experiment to prove that this isn't the case. They showed a video of basketball players passing the ball and asked people to count the number of passes made by the team wearing either the white or the black uniforms. In the middle of the video, one of two strange things happened. "Either a woman with an umbrella or a person in a gorilla costume unexpectedly walked through the center of the action, remaining clearly visible for about five seconds." The subjects of the experiment were asked afterward if they noticed anything odd. "Thirty-five percent of the observers failed to notice the woman with the umbrella, even though her presence was obvious to anyone not engaged in the counting task."

A control group of people who were not counting passes easily saw the woman with the umbrella because when they were asked to look at the screen, it was with the attitude: What's up? What might I see there? It was an attitude open to an unfamiliar world, accepting of whatever was there. There was no model and there were no expectations. The order "Tell me what you see" produces curiosity. The order "Count the passes" produces a closed system, a narrowing of attention directed at a particular task, which fills up working memory. The implicit assumption is that you know what you're doing and know what sort of perceptual input you want. However, the demands of the task use up scarce resources in the brain. It is a magic trick in that it creates a model that directs your attention away from something obvious that the magician (the researcher) is going to do right before your eyes. Magic confirms the idea that you see what you expect to see, and that under the right circumstances, working memory can't be distracted from its task. Such a closed attitude can prevent new perceptions from being incorporated into the model. Such a closed attitude can kill you.

Psychologists who study survivors of shipwrecks, plane crashes, natural disasters, and prison camps conclude that the most successful are open to the changing nature of their environment. They are curious to know what's up.

But what's most surprising in the experiment at Harvard is that an even larger number of people (56 percent) didn't notice the gorilla.

One researcher said that he'd ask the subjects, "Did you see anyone walking across the screen?" No, they had not. "Anything at all?" Nothing.

When asked, "Did you notice the gorilla?" the observers would say, "The *what?*"

Another researcher had commercial pilots fly a simulator down to a normal landing. Just as they were making the final approach to landing, the computer caused an airplane to appear where the pilots were supposed to land, parked right in the middle of the wind-shield. Many pilots just went right on landing without seeing it.

Everyone says that the mind plays tricks, but deep down, most people don't believe it. A brisk walk through city traffic will prove that we trust perception to represent the real world faithfully. Yet every day you experience numerous small events in which you lose your senses and then regain them. You say that the mind wanders, and it does, often farther than you think. While everyone is famil-iar with doing something stupid, it's not stupidity. Focusing atten-tion takes energy, and even then it can be fragmentary. (The word "focus," being a metaphor of a lens, can be misleading, since sometimes what's needed is a wider field of view.) The rest of the time, emotions and mental models of ourselves and our environ-ment must work automatically to create behavior. Most people who are apt to read this book live in low-risk environments designed specifically so that the consequences of inattention are trivial. But if they take that attitude into the wild—or into any risky area of life, love, or business—the cost can be high. Actually, the cost can be high anywhere. Recent research demonstrates that talking on a cell phone while driving, whether hands-free or not, causes inattentional blindness, the psychological term for the phe-nomen of failing to see the gorilla.

At first, it seems to make no sense that you could fail to see a gorilla; but if you understand mental models and working mem-ory, it makes perfect sense. You're counting how many times the players pass the ball. Fine. You set up a model—a simplified repre-sentation of ball, man, and motion—to facilitate that task. Now a gorilla appears. Gorillas are not helpful in completing the task. They have nothing to do with basketball. Gorillas are irrelevant and would displace the task in working memory. So the brain, effi-cient system that it is, filters out the gorilla so that you can keep counting. Seeing the gorilla would be a mistake. You'd lose count.

What's more worrisome is the pilot not seeing an aircraft on the runway, because that is relevant to the task he's concentrating on. But the model he created before the plane appeared did not contain the plane. And anyway the pilot wasn't landing a real plane (nor even a real simulator). He was landing a mental model of a plane on a mental model of a runway. (Simulators could not be used for training if it weren't for mental models.) Once the environment changed, the pilot's mental model was out of date. Some people update their models better than others. They're called survivors. I am not suggesting that the river guides didn't see the trees shooting down the river. I don't know. They're dead. The light certainly must have hit their retinas. But the research in perception and mental models makes it easier to imagine how they could have discounted their ultimate significance.

"In the face of uncertainty," Charles Perrow writes, "we must, of course, make a judgment, even if only a tentative and temporary one. Making a judgment means we create a 'mental model' of an expected universe. . . . You are actually creating a world that is congruent with your interpretation, even though it may be the wrong world."

FIVE

THE ANATOMY OF
AN ACT OF GOD

ON JUNE 25, 2000, two brothers, Rob and David Stone, and their friend, Steven Pinter, all in their twenties, set out to climb the southeast buttress of Cathedral Peak in Yosemite. Although it was rated as an easy climb (5.6), it involved six pitches (rope lengths) to an altitude of 10,940 feet. It was a serious undertaking. In addition, it was Rob's first climb of more than one pitch. But they were young and strong and well educated, David was more experienced, and if everything went just right, it would be fine. The annoying thing about plans is how rare it is for everything to go just right.

They had driven from the San Francisco Bay area to Yosemite and set up camp about fifty miles from the trail the night before the climb. All the while they were moving toward a goal: the Big Climb, where the plan, a memory of the future, tries on reality to see if it fits.

Sometimes an idea can drive action as powerfully as an emotion. Plans are an integral part of survival. Plans are generated as one of the many outputs of the brain as it goes about its business of mapping the body and the environment, along with the events taking place in both, resulting in adaptation. Planning is a deep instinct. Animals plan, and a bird that hides seeds has a larger hippocampus than others, suggesting a larger capacity for spatial memory. But planning—predicting the future—may be even more fundamental than animal abilities suggest. In his book *Complexity*, M. Mitchell Waldrop points out that "All complex adaptive systems anticipate the future. . . . Every living creature has an implicit prediction encoded in its genes . . . every complex adaptive system is constantly making predictions based on its various internal models of the world. . . . In fact, you can think of internal models as the building blocks of behavior. And like any other building blocks, they can be tested, refined, and rearranged as the system gains experience."

The human brain is particularly well suited to making complex plans that have an emotional component to drive motivation and behavior. Plans are stored in memory just as past events are. To the brain, the future is as real as the past. The difficulty begins when reality doesn't match the plan.

Memories are not emotion, and emotion is not memory, but the two work together. Mental models, which are stored in memory, are not emotions either. But they can be engaged with emotion, motivation, cognition, and memory. And since memories can exist in either the past or the future, to the brain it's the same thing. You bookmark the future in order to get there. It's a magic trick: You can slide through time to a world that does not yet exist.

David Stone, who had gone rock climbing numerous times, had a mental model of the world of rock walls. He understood the rules that governed them and his movement on them. He had a model of the systems he used, too, the ropes, anchors, harness, quick draws, carabiners, daisy chains—rock climbers tend to know their stuff. Those elements in his brain had no inherent emotional component, they were just schematics of that world.

He also had memories of his climbs on other cliffs, and the

excitement he experienced gave those memories a nice emotional quality. Just thinking about climbing again, remembering the future, felt good. It gave him that "beacon of incentive." Without having to think about it, he took his mental model of cliffs and superimposed it on the Cathedral Peak, a face he'd never climbed. Conditions had been good in the past, the climbs were successful, the feelings good. So that was the memory, the model, he applied to Cathedral Peak. Logically, he would have been able to tell you, had you asked, that he had no way of knowing what climbing Cathedral Peak would be like. It was a new world, but he was approaching it as if it were familiar. How else can we move forward in time? It's the risk-reward loop.

Like David, we all make powerful models of the future. The world we imagine seems as real as the ones we've experienced. We suffuse the model with the emotional values of past realities. And in the thrall of that vision (call it "the plan," writ large), we go forth and take action. If things don't go according to the plan, revising such a robust model may be difficult. In an environment that has high objective hazards, the longer it takes to dislodge the imagined world in favor of the real one, the greater the risk. In nature, adaptation is important; the plan is not. It's a Zen thing. We must plan. But we must be able to let go of the plan, too.

PSYCHOLOGISTS WHO study survival say that people who are rule followers don't do as well as those who are of independent mind and spirit. When a patient is told that he has six months to live, he has two choices: to accept the news and die, or to rebel and live. People who survive cancer in the face of such a diagnosis are notorious. The medical staff observes that they are "bad patients," unruly, troublesome. They don't follow directions. They question everything. They're annoying. They're survivors. The *Tao Te Ching* says:

The rigid person is a disciple of death;
The soft, supple, and delicate are lovers of life.

My father was a rule breaker, as I have turned out to be. He broke nearly every bone in his body in the war and then broke the rule that it's supposed to kill you. He also wore a big beard when Castro took power in Cuba, partly to cover the scars, but also because it freaked people out. Beards were not cool in 1959; they were an emblem of communism.

The plan, then, can become the equivalent of doctor's orders, the tyrannical rule you can either obey or rebel against. It's one of the many delicate balancing acts of nature: to plan and not to plan. But emotion is a dramatic arc, and like a play or a movie, like sex or the hunt, it has to run through three acts, with a struggle in the middle and a climax near the end, when our eyes grow wide. Rob and David Stone and Steven Pinter had fallen under the tyranny of the rule created by their own plan and were quite naturally following it.

The closer they got to the goal, the harder they tried, the more excited they became. Halfway through a Hollywood movie, the hero becomes totally committed to his goal. So there they were, set on that natural emotional arc, and the plan gradually became unshakable. The overwhelming evidence that conditions in the real world were rapidly diverging from their memories of the future made no difference—after all, in a classical drama, the tension comes from the fact that the hero is farthest from his goal when he goes on to triumph. (Let's not forget that, in some movies, he dies.)

As additional stresses or unplanned events make it difficult to think clearly, it's no wonder that you find it almost hopeless to change the plan or alter your model to fit reality. You're carried along on your own story of yourself.

WHEN THE THREE CLIMBERS awoke at 4:00 A.M. to begin their approach, the first unplanned event occurred. Someone ("not a bear") had stolen their food. The plan called for a hike of three miles to Cathedral Peak and a starting time of eight o'clock. David's calculation of how long the six pitches would take put them at the summit around one in the afternoon. But now they had to go in search of food. It delayed them for two hours, and another stress was added to being tired and hungry: the pressure of schedule.

With luck, they calculated that they still might make it to the summit by three . . . Hell, they had to. They'd driven all this way. The friction, made up of numerous minor events, was building, and they would try to overcome it.

They could have acknowledged the friction in the system and lived with it. There were lesser routes to climb that were closer to the road. They could have had a perfectly good day. But the plan put them on a three-mile hike, which put them at the base of Cathedral Peak, and that was that.

So they rushed over to check the weather board. The sun had come up. The day was gorgeous. But they were just being responsible, doing what they'd done before. The weather board still had the previous day's forecast on it. It called for clear weather, which predicted exactly what they could see for themselves. They knew it was outdated, but why should they expect it to change? They saw what they wanted to see and disregarded what they knew: that high terrain makes for rapid changes in the weather. In the summer, the middle of the afternoon is the most likely time for orographic lifting of warm, moist air, causing weather on those peaks at around 3:00 P.M., precisely when they'd be arriving.

Researchers point out that people tend to take any information as confirmation of their mental models. We are by nature optimists, if optimism means that we believe we see the world as it is. And under the influence of a plan, it's easy to see what we want to see. As for the outdated weather board, "That was good enough for us," as Rob commented later. They could have gone to the ranger station, but by now they were in a hurry. They could have brought

a weather radio, too. They could have brought waterproof clothing, but the sun was out. Trivial events begin to shape an accident
long before it happens.

When they arrived at Cathedral Peak, two other climbers were
already underway: Lewis Johnson and Stuart Robertson. So Rob,
David, and Steven got ready, but it was ten o'clock before they
started up. Two hours doesn't seem like much. But two pitches
later, they could see the clouds moving in. They kept climbing.
They'd been awake six hours. Also, every foot of altitude brought
less oxygen to their brains. Things were getting complicated. The
stress load was growing. Friction was building. And unfortunately,
they had made no firm bailout plan to tell them when to turn back
once stress had begun to erode their ability to reason.

Despite the fact that "by the end of the third pitch, with three
more to go, we could see distant rain," they all felt that "we were
outrunning the storm." They actually discussed the weather and
"made a group decision to press on for the top instead of rap-
pelling off." Even if they had succeeded, they did not consider
how rapidly hypothermia could overtake them in their cotton
clothing in a cold rain. They were locked in a game of speed chess
with Mother Nature. And she unleashed a series of stunning moves.

The hailstorm began as they were on the last pitch. David, who
was leading, had managed to reach the peak. Rob and Steven were
just about to begin the last pitch with David belaying from above,
when suddenly they all felt their hair stand up. Rob knew about
St. Elmo's fire, as sailors call it. As the negatively charged lower
portion of a thunderhead passes over the ground, it attracts a positive charge from the earth. The charge moves along under the
cloud and follows it. When the positive charge reaches anything
that can conduct electricity, such as a person, it moves up through
the person and creates what's known as a corona discharge. That's
what Rob and Steven were feeling now as they headed up the sixth
and last pitch.

Just before they reached the summit, as Rob later said, "everything around us started to buzz. That was the most terrifying sound I had ever heard!" Now there was only one emotion, and it blotted out everything else. Heedless of the risk of falling, Rob and Steven charged across the wet surface to an overhanging rock in an effort to escape the lightning. They had not reached the summit but were still clinging to the nearly vertical wall. As they waited beneath a Bible-black sky, the storm slashed at them with hail and rain, and the "crackling buzz continued for over five minutes."

Then it happened: Checkmate.

The detonation hammered Rob into the stone wall. He felt the electricity surge through him and out his right arm. It was over in less than a hundredth of a second (lightning can travel as fast as 54 yards per microsecond). Then he heard Steven moaning beside him. Lewis Johnson and Stuart Robertson, who were on the summit with David, were screaming.

Rob, who had almost reached the top, was near enough to his brother to see him dangling limply from his belay position on the summit. David was moaning weakly. The sight of another person who is injured or dead, especially someone you know, can elicit powerful emotional changes, well known among search and rescue workers: rescue fever. It's a common human reaction to do something foolhardy and even life-threatening while trying to help others. When the World Trade Center collapsed, a man tried to swim from New Jersey to Manhattan to help. He was picked up by a ferryboat captain.

Rob found himself scrambling without protection across the wet, extremely exposed rock to save his brother. Despite the risky position he was in, he began rescue-breathing. Any mistake on the exposed slab would have been fatal. But he could smell David's burned flesh, and the sensory input only heightened his emotional state, a sort of emotional shock that put him in a position where he "could easily have become a second victim." His cognitive abilities were all but shut down, as is typical in such extreme cases. It was only much later that Rob would consider the risk he had taken.

So there they were, sticking up into a black storm cloud on a spindle of rock, four guys with an injured man down and possibly dead, trying to move his sagging, 190-pound frame away from the edge and going hypothermic now because they were all wearing what the YOSAR (Yosemite search and rescue) crew call "death cloth"—cotton. They had no waterproof clothing, nothing with which to start a fire, no way to warm David or themselves. The hour was getting late, and they had only one flashlight among them. How different the world seemed now from those memories of the future.

As Johnson and Robertson tried to move David, he awoke, believing that he was in British Columbia. Johnson and Robertson set up a rope on an anchor point to lower him. While they were working, the buzzing returned: St. Elmo's fire. Rob and Steven had already developed a powerful secondary emotion and reacted without thinking. They jumped onto David's rope, putting the weight of four people on the tiny anchor all at once. Fortunately, their luck (and the anchor) held. It was also fortunate that Johnson had a ham radio and was able to reach someone, who contacted John Dill, the YOSAR ranger. Had they waited to see the updated weather board that morning, they would have noticed warnings of lightning. Mother Nature must have been feeling good that day: They all lived. As the accident report noted, "Mother Nature chose her targets with a sense of humor: Four are electrical engineers."

IF YOU distill all of the psychology, cognitive science, and neuroscience of the last hundred years or so, what you find is that we're always *Homo* but sometimes not so *sapiens*. People are emotional creatures, which is to say, physical creatures. Joseph LeDoux concluded that, "people normally do all sorts of things for reasons that they are not consciously aware of . . . and that one of the main jobs of consciousness is to keep our life tied together into a coher-

ent story, a self-concept." In other words, everyone is the hero in
his own movie.

So it should come as no surprise that, in many cases, basic survival mechanisms, which have been hardwired into us and
sculpted by experience, turn out to be not only the most powerful
motivators of behavior but to operate at their peak efficiency out
of reach of the conscious decision-making powers, which makes it
easy for reason to be overwhelmed. Once an emotional reaction is
underway, we can be swept away by an irresistible impulse to act.

But there are many ways of revising the script and adapting in
hazardous situations. Training is one of them. Neil Armstrong
would not have been able to land *Eagle* on the moon if it hadn't
been for years of rigorous training, not only in the technical stuff,
which had to be second nature, but in emotional control. All elite
performers train hard, and when you follow in their path, you'd
better train hard, too, or be exceptionally alert. That's the main
difficulty with neophytes who go into the wilderness: We face the
same challenges the experts face. Nature doesn't adjust to our level
of skill.

The practice of Zen teaches that it is impossible to add anything more to a cup that is already full. If you pour in more tea, it
simply spills over and is wasted. The same is true of the mind. A
closed attitude, an attitude that says, "I already know," may cause
you to miss important information. Zen teaches openness. Survival
instructors refer to that quality of openness as "humility." In my
experience, elite performers, such as high-angle rescue professionals, who risk their lives to save others, have an exceptional balance
of boldness and humility. So do astronauts.

Just being aware of the traps that can be set by nature can help.
It helps to remember that people are primates, with a recent and
somewhat untested upgrade, the neocortex (called "neo" because
it was thought to be new; what's new is its size and complexity).
What we regard as failures of mind are probably nothing more

than the process of nature tinkering with simple rules over a leisurely span of evolutionary time. Nature always runs through many individuals of every species in its experiments, and we are the latest experiment. It's nothing personal, then, when the brain plays tricks. Nothing personal, either, when you die, as Marcus Aurelius, the Roman emperor and Stoic philosopher, said.

Our failures are so common that it's easy to write them off as inexperience, stupidity, or inattention. Most people operate in an environment of such low risk that action, inaction, or the vicissitudes of brains have few consequences. The energy levels, the objective risks, are low. Mistakes spend themselves harmlessly and die out unnoticed instead of growing out of control. Even so, other than on America's highways, which see more than 45,000 fatalities each year, most accidental deaths happen in and around the home, largely because that is where people happen to be while making routine mistakes. But when you take yourself out of that environment and go into the wild, or when you face a new challenge in everyday life, you must evolve new ways of seeing, a new plan.

A FEW years ago, Lyle Lovett and I went down to Ensenada on Chris Haines's "Baja Off Road Tours" with a group of twenty guys, and we rode five hundred hard miles in the desert and the mountains. It was the sixth time Lyle had taken the trip with Chris, a motorcycle racer who had run the Baja 1000 for fifteen consecutive years. In the liner notes on Lyle's album *The Road to Ensenada*, it says, "Thanks to Chris Haines, Mike's Sky Rancho, the San Nicolas Resort Hotel, and Papas and Beer," which is more or less a map of the trip we took. The group included Lyle's father, Bill Lovett, to whom *The Road to Ensenada* is dedicated, as well as his assistant, Vance Knowles, and his cellist, John Hagen. Percussionist James Gilmer drove one of the support vehicles, and Lyle's mom, Bernell, rode shotgun in it. Lyle also brought some motorcycling buddies, who ran the spectrum from David Bouley, a chef from New York, to Bill Kasson, owner of a motorcycle shop in Austin.

The ride began on a dirt road that rapidly deteriorated as it was cut and crisscrossed by erosion from torrential rains, which had guttered the earth. The cuts turned to ditches, which turned to ravines, and before the first hour was out we had a casualty, when a motorcycle shop owner from out east came into a corner carrying too much speed and stuffed his machine into a deep ditch. (We call that "taking a soil sample.") I came around the corner to find him standing in the middle of the path with his face bloody and his thumb broken or at least badly bent. We were a long way from medical attention. He got into the truck with Bernell. His ride was over.

The first day began at the San Nicolas Hotel and ended at Mike's Sky Rancho, a famous Baja motorcycle hangout high in the mountains, with precious little electricity, no phone at all, and an ancient swimming pool that looked like a poisoned waterhole. By the time we had all made it up the mountain and beer had been rationed out, cigar smoke was wafting across the crowd by the pool, mingling with the smell of tortillas from the kitchen. The sun was going down as we realized that Lyle's father, Bill, was missing.

Lyle, his mother, and I walked into the parking lot and looked out at the mountains, as if they could tell us something of his fate, but the sun failed and the immense shapes of the hills melted into the moonless sky, and still there was no sound of a Kawasaki four-stroke engine out there among the pines. The air grew cold. Many of us had taken ice-cold showers in the dim, four-bunk rooms and put on sweaters and jackets, and now we stood around, hugging ourselves, thinking about Bill out there, alone. Everyone else was accounted for, and it was bad, because we weren't looking out for each other, and we were ashamed. After all, if we don't look out for each other, we're no better than coyotes.

An hour after dark, Chris mounted a search party consisting of two men on bikes and one in a truck. They encountered Bill, half frozen, doggedly making his way up the mountain inch by inch, following the faint glow of his Kawasaki 250's trembling headlight.

At dinner, Lyle and Bernell and Bill sat together in the big loud room with more than sixty bikers eating and drinking and whooping it up, and Bill swore he'd never get on a motorcycle again. But then he said, "Well, we don't like to make decisions at night in our family." As he ate, Lyle kept leaning over to put his hand on his father's shoulder and saying softly, "You all right, Daddy?" Bill would nod and say yes, yes, he was fine, and he ate a big dinner of beans, tortillas, salsa, and chicken, then he and Bernell got up from the table to turn in early. When he said good night to his son, Bill leaned over and kissed him on the cheek. Lyle and I stayed a little longer at the table, Lyle eating hot tortillas rolled up with sugar, which he poured out in a white cascade from the glass sugar shaker.

In the morning, Bill was ready to go again. He and I rode together, vying for last place, through the next two days of goat track and sandwash. Lyle went fast and we went slow, and although Bill and I got lost one more time (in a tomato farm on the Pacific sand during a windstorm), we managed to find our way out again; and while a lot of the riders fell, neither Lyle nor Bill nor I did. David Bouley tore the front fork off his 650 to avoid killing a rider who had tried to swim with more chain than he could carry and fell in David's path.

At midday we entered a vast dry lakebed as hard and flat as poured concrete and dusted with alkali powder as fine and white as confectioner's sugar. It was a chance for everyone to go as fast as they pleased, and we all took off in a towering cloud of dust that rose and moved across the land. I was the slowest one, as even Bill left me behind, and I quickly lost sight of them. But as I gained more confidence and speed, I decided to catch up. I nudged the speedometer above 100, then 110, then gradually began to see their dust cloud again and turned toward it. It was impossible to see tracks on the lakebed, so I had only a general idea of which direction to go.

When the cloud of dust grew, so did my confidence, and I cranked the throttle all the way open and decided it would be best not to look at the speedometer again, which might scare me. Now the cloud was getting huge, and I could picture myself pulling up behind a group of green motorcycles at any moment, maybe even passing Bill. Then suddenly, out of the cloud, there appeared a battered red Volkswagen bus full of Mexicans, and at the last second, before the impact, I swerved off to one side, lost control of the bike, and skidded around until I became mired in a dry sandy wash. The bus went roaring away, dragging its skirts of dust. The whole thing had been an illusion, and I'd barely avoided a bad accident because of it.

The third day we were flying up a dry silt road that gradually ascended the hills above that same dry lakebed. A dirty wind was blowing dust all over the valley, obscuring the mountains and making the lakebed impassable. I was going about as fast as I dared, feeling my bike's rear end skip and skid out from under me as I powered through, when the landscape, the jagged rocks and the alkali lake in the windstorm, and the high blue of the sky overhead all at once came together into a coherent picture. I became one with the terrain and the motion. I saw the fine fabric that the wind had woven out of sand and rock, and I understood the transformation that was taking place through this, the simplest of all machines: the wheel. The essence was this: When it was still, that wheel was merely a contraption of spokes and nuts, of rim and rubber, cobbled together by human hands, improbable as a saxophone. But by spinning, it became a singular jewel, a unity of glittering perfection, fashioned out of higher laws, transformed by wind just as the saxophone is when it is played.

So with the land we covered. Now, as we accelerated above it, touching it but lightly, it was soft as angora. But if my motion was disturbed, then this delicious frothy blanket would erupt into a thousand shards of rock and scalding sand, and all would resolve at once into the harsh reality of this lifeless lake. That's what falling was all about. Riding was transcendental; it was rolling the karmic wheel in order to ascend with angels out of the temporal hell of the flesh. Falling was to reenter the world as it was, the low world. Falling, we were all fallen angels. Hence, the treachery of our expedition was in essence the same as Satan's. Early bikers understood this concept and had named themselves accordingly.

SIX

THE SAND PILE EFFECT

AT ABOUT TEN minutes to nine on Thursday morning, May 30, 2002, Bill Ward sat at the top of Hogsback Ridge on a natural shelf just below the summit of Mount Hood in Oregon. His team of four climbers had begun its descent. Chris Kern was a forty-year-old investigator for the New York appellate court. Harry Slutter was his old friend from Long Island. Slutter had met Ward, an experienced climber, through his job with a large commercial nursery that had offices in New York and Oregon. Ward, in turn, had brought along his friend Richard (Rick) Read, a local Oregonian who'd never climbed a mountain before.

Read, who would be dead within the hour, had been led to believe that Mount Hood was a beginner's mountain, suitable for his first climb. Unfortunately, there's no such thing as a "beginner's mountain." It's a concept that doesn't work, like beginner sex. One difficulty is that the standard route up (and down) Mount Hood is not technical. It's more of a hike on a steep snow field. On a good day, you can walk it without crampons, snap some pictures on the summit, and be back at the Timberline Lodge ski resort for

New Zealand fire-broiled spicy lamb loin chops. It's a dangerous illusion, because success depends on doing everything perfectly. Any fall is apt to be a very long one into inhospitable terrain. Mount Hood is an active volcano, with glaciers, ice fields, sudden 140-mile-an-hour winds, and rime ice. The fumarole, a volcanic vent, sucks in those who fall and suffocates them with hydrogen sulfide gas. People slip and accelerate across the ice, slam into the Steel Wall, and die on the rocks. Sudden whiteout blizzards leave people wandering for days in the Zig-zag Wilderness Area or the crevasse field on the Eliot Glacier. Mount Hood claims at least one life each year. Some years, as in 2002, it claims a lot more.

Starting at about eight-thirty in the morning, Bill Ward and his crew had descended from the summit and gathered on the shelf above the steep Hogsback Ridge, a graceful catenary arc of ice that descends 1,000 feet from the summit. Kern put his ax into the snow as an anchor and belayed Slutter. He made sure that there was no slack as he fed the rope out to Slutter for his descent. If Slutter fell, the taut rope would stop him before he got going too fast. Recent rains—freezing at night, partly thawing by day—had turned the snow into a mixture of hard ice and soft slush, making it easy to fall but difficult to stop.

They had practiced self-arrest techniques in the days leading up to the climb. Everyone, including Rick Read, the novice, had been able to stop. When one person on a team falls, he's supposed to shout, "Falling!" and all the others are supposed to throw themselves down and bury their ice axes in the snow. The rope tying them together should arrest the fall. With experience and practice, it becomes second nature, a secondary emotion, such as Remarque's soldiers had developed in response to the whistling of high-explosive shells. But although it had worked in practice, belaying from a fixed anchor is more reliable.

Once Slutter had descended 35 feet, Ward put his ice ax in, wrapped the rope around it, and belayed Kern. Kern pulled his ice ax out of the snow and started down. As he moved, so did Slutter, in order to keep the rope taut.

The plan, the idea, was that everyone would move down on belay one at a time until there was no slack in the rope. Anyone who fell would fall only a few inches. Then, lined up like beads on a string on the 1,000-foot ridge, they'd pull the last ice ax and walk carefully down the mountain.

Ward and Read watched from the ledge until the rope was taut. Then Ward belayed Read, who descended his 35 feet, taking up the slack between himself and Ward. Slutter and Kern moved, too, to keep the rope taut.

When they'd played out all their rope, Slutter was 105 feet below the shelf, Kern was 70 feet down, and Read was 35 feet below Ward, who still sat on the shelf with his ice ax in, protecting the team. Now Ward stood and pulled his ice ax from the snow and the only thing holding them to the mountain were the crampon points. The system was free to behave as it might.

It was a simple theory, and from my interviews of the survivors, it was clear that they'd given it a lot of thought. The difficulty with the theory was that the top man, Ward, must not fall. (The order from top to bottom was Ward, Read, Kern, and Slutter.) If you drop a brick 6 inches, you can probably catch it safely. If you drop a brick out a third-story window, it could hurt somebody. There were 35 feet of rope between Ward and the second man, Read. If Ward fell, he'd have to go 70 feet before the slack was gone. Think about catching someone who's fallen from a seven-story building. It's all about energy levels, and while they didn't put it this way, the climbers were attempting to manage the vast amounts of energy they had put into a system of rope and weights. Only at an instinctive level did they understand that it was critical to keep that energy from escaping all at once. Ward, being the most experienced, was at the top. Surely, he wasn't going to be the one to fall. Everyone was worried about Read. The word "experienced" often refers to someone who's gotten away with doing the wrong thing more frequently than you have.

By roping themselves together, they had created a deceptive system. A rope is simple, yet capable of surprisingly complex behavior. It can transmit all the force imparted to it from an infinite number of points along its length. It can double and double again. It can transmit force along its length and deliver it somewhere else. It can stretch, shrink, vibrate, and break. It is such an elegant equation. But powered by the infinite resource of gravity and coupled to four bodies, it turns out to be capable of a staggeringly complex array of outcomes.

Slutter, the lowest man on the rope, was looking down at some other climbers. The two nearest ones had celebrated with him and his team just half an hour earlier, snapping photos, laughing, sharing water and candy bars on the summit. Slutter could see them now: John Biggs and Tom Hillman, both from California. Hillman was a Methodist pastor, Biggs was his parishioner.

"I was making a mental note at the time of where Hillman and Biggs were," Slutter told me a few days later. They were off to one side, so he assumed that they were not in his fall line. It was one of the many illusions of Mount Hood. Slutter turned to look over his right shoulder at Kern, 35 feet above him. "I had a better view of where the spine [of the Hogsback] went." So he called up to Kern, "Back up a little and get yourself over the spine more."

Then, out of the corner of his eye, Slutter saw a blur and felt a tremendous surge of norepinephrine. He believed that whoever was falling could not be from his group, because the blur was going to one side of him and not straight down toward the Timberline Lodge. Two days after the accident, I made the same mistake of perception myself. I was climbing the mountain with Steve Kruse, the mountain manager for the ski resort. He asked me which way was down and I pointed at the lodge. But a ball rolled from the top does not go to the lodge. It goes east into the Zig-zag Wilderness or west into the Eliot Glacier.

Nevertheless, appreciating the precariousness of his position, Slutter reacted instinctively. "There was no hesitation," he said. He threw himself down and dug in with his ice ax.

Moments before, another group of four climbers had appeared above Ward from the summit just in time to see what happened. The criminologist for Clackamas County, Tim Bailey, investigated the accident. In his report, he said that the entire fall took only three to five seconds, covering hundreds of feet on the way to a crevasse that cuts the Hogsback Ridge about halfway down. A teenager named Luke Pennington was in the group of four that had appeared above the victims just as Ward pulled his protection. He saw Ward facing east, trying to turn and plant his left boot while descending. Then the one thing that must not happen did happen: The top man fell. "At that point he slipped and fell," Bailey said. "When he slipped and fell he landed on his back with his head facing down the mountain and he started sliding downhill."

Before his rope went taut against Read's harness, Ward was going as much as 30 miles an hour, the equivalent of the speed you'd attain by jumping from an eight-story building. Once he took Read (then Kern, then Slutter), the acceleration continued unabated. They had taken to the boundaries of their world for adventure. And as James Gleick writes in *Chaos*, "Strange things happened near the boundaries."

With all that energy balanced on ice on a crampon point, it took only a slight tug, the pinprick that popped the balloon. "It played out in slow motion," Slutter said. The speeding up of reflexes, the fast-processing mode into which the brain and body go in an emergency, makes the world seem to run in slow motion. "I know that Chris went into the arrest position, but I don't remember seeing him move. It's amazing because I didn't hear a word." Perhaps it was a result of perceptual narrowing. Perhaps no one said anything. "I saw nothing else. I was facing up the hill. I remember looking at my ax dug in, and my chest was on top of it. My head was cocked and my hand was over it." Slutter focused on the one thing that seemed most important: his anchor point. "Then I remember watching it ripping through the ice. I'm thinking: We're going pretty fast here. I was ripped from the mountain."

Kern was ripped from the mountain, too, and remembers flying through the air. He couldn't tell how far, but it seemed like a very long way. When he hit the snow again, it was a stunning blow that broke his pelvis. Then, as the ropes began playing out and going taut, he was snapped free again, flung into the air, and sent rocketing down the mountain.

Perhaps only a second had passed. Perhaps two. All four were now in chaotic motion, accelerating toward Biggs and Hillman, who were carefully descending. They were less than 100 feet below. They had no chance.

Bailey reported: "Mr. Pennington said when they reached the next group . . . they clotheslined Mr. Biggs. The four falling were falling in such a manner that the taut rope was stretched out and it hit Mr. Biggs, it actually knocked Mr. Biggs into the air."

"I heard someone shout 'falling!'" Hillman recounted. "They hit Biggs like a billiard ball. He was airborne for three or four feet and was knocked upside down. I saw the one Oregon guy hook his red rope on our blue rope as he passed. I knew that I would have to arrest five people, and I stopped watching and got down. I prayed because I knew that I would have 50 feet until they reached me," which was the length of rope between him and Biggs, "and then 50 feet more until I was engaged. I was ready." But with 100 feet of slack, at least a couple of the five bodies, maybe more, would fall ten stories before they pulled Hillman's rope taut. By that time, the force was so great—thousands of pounds—that it ripped his shoulder from its socket. Still, he held on. "I remember the ax ripping the ice all the way down until the crevasse."

But it wasn't over yet. Just above the crevasse, perhaps 400 feet below, two teams of climbers were heading up the mountain. They were led by Jeff Pierce, a paramedic with the Fire and Rescue Department in nearby Tualatin Valley. He'd brought four

other paramedics up for their first experience of climbing a mountain. In addition, the department's physical fitness trainer, Chad
Hashburger, and Cole Joiner, the fourteen-year-old son of one of
the paramedics, had come along. After their 3:00 A.M. Snow-Cat
ride to the top of Palmer, the highest ski lift, Pierce had divided
them into two groups. He led Cole Joiner and Jeremiah Moffitt, a
paramedic, on the lead rope. The other four formed the second
team, led by Dennis Butler, also a paramedic, and the only other
experienced climber among them. His team of four was still below
the crevasse when the accident happened.

Pierce had already led Joiner and Moffitt around the left side of
the crevasse as they ascended. "I was coming around the crevasse,"
he remembered, "looking up at some other climbers high on the
route above, and planning my moves." Pierce wanted to stay out of
the way in case someone fell, but because of a seemingly insignificant decision he'd made earlier (to go to the left instead of the
right), his team first had to cross beneath the higher climbers. He
and Joiner then started moving to their right. But Moffitt was still
in the fall line.

Pierce looked up and saw the teams falling. He'd seen people
fall before; therefore, as he told me, "I expected them to drop, self-
arrest, and start descending again." But as the first rough images
began to resolve in his brain, a tremendous jolt of chemicals
flooded his system and set him in motion. "I dropped and buried
my ice ax. I really dug in, and everyone else did, too." Six human
projectiles, tangled in a mess of ropes and bristling with the points
of crampons and ice axes, were plummeting toward him.

Bailey reported: "Now we have six climbers coming down. The
other witnesses . . . said by this time they were all bouncing off
each other, they were spinning around in circles."

Moffitt was hit so hard he was knocked unconscious. His rope
jerked Pierce and Joiner into the crevasse, where all nine climbers
hit the lower wall and settled into heaps of bodies. Biggs was
already dead. Ward was buried head-down in the snow, suffocating

under a pile of bodies. Read, on his first and last climb, was alive, conscious, and dying.

At that moment, the manager, Steve Kruse, was opening the mountain for the day's skiing. He was at the top of the Palmer chair lift talking to a ski patroller when the boy in the lift house said there was a call for him. Kruse heard a sheriff's sergeant tell him there'd been a 911 call reporting a climbing accident 2,000 feet above him at the bergschrund, which is what the locals call the crevasse. It was a little garbled, a little wild, but it sounded as if a bunch of climbers had gone into the 'schrund.

Kruse's head snapped around. Far above, he could see the crevasse with a few small figures around it. He felt his heart tick into a canter. He knew this was going to be a big one. He grabbed his radio, called down to Jeff (Floodo) Flood, "my ace Cat driver," and told him to get up to the top of Palmer as fast as he could. He could see the yellow Sno-Cat snort black diesel smoke as it began churning uphill from the Timberline Lodge.

ABOVE HIM, deep in the crevasse, Bill Ward, buried in the snow, had already suffocated. Chris Kern was folded in half, jammed beneath a rock, with a broken pelvis. He was screaming in agony. Harry Slutter, who'd been leading the highest team down, had been knocked unconscious when he hit the crevasse wall. He awoke jammed upside down, his ice ax still in his hand. The scene, lit by ice-blue sunlight, was surreal. Debris and snow were still pouring in from above. "I felt like I was drowning, breathing in all that fine snow," Slutter remembered.

He righted himself and took stock of his injuries. He was almost certain that he'd broken both his jaw and an ankle, but he was more concerned about Biggs. "I think my body was on his or his on mine; there was no separation," Slutter said. He rolled over and checked: Biggs wasn't breathing. Slutter began trying to resus-

citate him. Then Pierce, leader of the Tualatin group, came over and "he confirmed what I suspected," Slutter said. "John was dead."

Kern was still screaming. Slutter shouted at him: "Put away the pain and hold on!" He and Kern ran races together, and that's what they told each other when the going got tough. Kern quieted down and tried telling Rick Read the same thing. Read was talking as he died. Moffitt was moaning, babbling incoherently.

Tom Hillman, the Methodist pastor, had been knocked unconscious, but his Camelback water bag had burst against the wall, absorbing some of the impact. Despite that, he'd cracked a thoracic vertebra. Hillman, who was trained as an EMT, began assessing himself as soon as he awoke. "I went back to my EMT training and before I moved, I did a mental evaluation. I thought at least I would have a tib/fib fracture, an ankle, or a femur. I was mentally ready. But when I woke up, I was amazed that I had not broken anything. I had torn muscle and ligaments in my arms, shoulders, and back from holding onto the ice ax in arrest. I wanted to be able to respond and go into rescue mode, but with my concussion I was like molasses." It is typical of the best survivors that, despite his injuries, Hillman was surrounded, as the *Tao Te Ching* puts it, "with a bulwark of compassion." His thoughts were not for himself but for others. It is not insignificant that he had chosen to devote his life to the ministry. "The hardest moment," Hillman said, "was not being able to respond with the equipment and training I had."

Every time Hillman moved, he brushed ice onto Kern, who would cry out in pain. Hillman crawled out of the way and lay down. Only Jeff Pierce and the fourteen-year-old Cole Joiner were nearly uninjured. They immediately set about helping the others. Dennis Butler, leader of the other team, which was still outside below the crevasse, began setting up a rope system to haul out the dead and injured.

. . .

THE TYPE of accident that happened on Mount Hood on May 30, 2002, was going to happen, as it always does, to someone somewhere. All the available theory tells us that it is an inevitable part of the larger system that puts climbers on steep snowy slopes in large numbers.

Climbers travel in roped teams without fixed protection all the time. They get away with it, too. They use ice axes like walking canes on descent. All the elements of the system in the Mount Hood accident were normal and can be explained by normal accident theory. It was Charles Perrow who coined the term "system accidents" in the 1980s, and it's a fascinating (if academic) exercise to see how his work and both chaos theory and the theory of self-organized criticality dovetail. The Mount Hood accident involved two broad categories of effects: the mechanical system that the climbers were using, and the psychology and physiology that contributed to the accident.

In recent years, those who study accidents in outdoor recreation have begun to recognize that all accidents are alike in fundamental ways. If you find yourself in enough trouble to be staring death in the face, you've gotten there by a well-worn path. Your first reaction might be: *How could this have happened? What rotten luck!* But if you are alive afterward and bother to examine what happened, it will all seem as orderly as the Cajun Two-Step.

The events that we call "accidents" do not just happen. There is not some vector of pain that causes them. People have to assemble the systems that make them happen. Even then, nothing may happen for a long time. That is how mountains such as Hood, McKinley, Longs Peak, and others get a reputation as milk runs. Many of the people who get into the worst trouble on such nontechnical peaks are those who have climbed more difficult mountains elsewhere. Often they have climbed in the Himalaya or South America and come to Denali or Hood with an attitude that they're slumming. Perhaps

they're doing a favor for a friend who wants to have the experience. Perhaps they're trying to climb the highest peak in every state. They are hijacked by their own experience combined with ignorance of the true nature of what they're attempting to do.

Perrow's *Normal Accidents*, first published in 1984, is a work of seminal importance because of its unusual thesis: That in certain kinds of systems, large accidents, though rare, are both inevitable and normal. The accidents are a characteristic of the system itself, he says. His book was even more controversial because he found that efforts to make those systems safer, especially by technological means, made the systems more complex and therefore more prone to accidents.

In system accidents, unexpected interactions of forces and components arise naturally out of the complexity of the system. Such accidents are made up of conditions, judgments, and acts or events that would be inconsequential by themselves. Unless they are coupled in just the right way and with just the right timing, they pass unnoticed. Bill Ward had slipped and caught himself before or was in a position where slipping didn't matter. He'd pulled his protection, too, but not just before a serious fall. Perrow's point is that, most of the time, nothing serious happens, which makes it more difficult for the operators of the system (climbers, in this case). They begin to believe that the orderly behavior they see is the only possible state of the system. Then, at the critical boundaries in time and space, the components and forces interact in unexpected ways, with catastrophic results.

The space shuttle had flown many times without incident. It had gone up on cold Florida mornings and nothing bad had happened. The people responsible for it began to regard this outcome as the only possible behavior of the system. Then, one morning, the cold caused a rubber seal to crack. The engineers who understood the system better than those in charge had warned about the possibility. Under the pressure of schedule and politics, *Challenger* was launched anyway, and those in charge, along with the entire world, saw a graphic demonstration of another possible, if rare, behavior of the system known as the space shuttle.

Perrow used technical terms to describe those systems. He called them "tightly coupled." He said that they must be capable of producing unintended complex interactions among components and forces. In his view, unless the system is both tightly coupled and able to produce such interactions, no system accident can happen (though other failures happen all the time).

The parts and forces and their potential interactions might be hidden and are difficult to imagine beforehand. The climbers could not see all the energy they had stored. Because they had watched self-arrest techniques work before, they couldn't imagine them not working again. The coffeemaker or the toilet on an airliner is not supposed to be able to destroy the plane, but both have done so.

An airliner is a perfect example of a complex, tightly coupled system: a large mass containing explosive fuel, flying at high speeds, and operating along a fine boundary between stability and instability. Small forces can upset it, causing the destructive release of the large amount of energy stored in the system. In that way it is like the system set up by the climbers: The energy in their system had come from their own muscles, electrochemical energy produced as they climbed. As they moved up, they stored more and more as potential energy in the system, which was tightly coupled because of the rope. It was like blowing up a balloon. The tiniest pinprick, an almost imperceptible force, can release the air all at once. The climbers would have been better off if they had tried to descend the slope with no safety system at all.

When a system is tightly coupled, the effects spread. In a loosely coupled system, effects do not spread to other parts of the system. Falling dominoes are a familiar illustration of how tight coupling works. Move the dominoes farther apart and knock one over: only one will fall. If the climbers had not been roped together, Ward wouldn't have taken everyone else with him. (For that matter, if they had not misperceived which way was down, they might not have positioned themselves over Hillman and Biggs.) But the accident was still no one's fault. There is no cause for such system accidents in the traditional sense, no blame, as the *I Ching* says. The cause is in the nature of the system. It's self-organizing.

WHEN *NORMAL ACCIDENTS* was published, neither chaos theory nor the theory of self-organizing systems was widely known or accepted. But it is possible to see hints of both in Perrow's work. Chaos theory arose out of a huge vacuum in the physical sciences: disorder. We see it everywhere we look, from the functioning of a living organism to the behavior of flowing water: turbulence; erratic behavior; nonperiodic natural cycles from weather to animal populations. Classical physics ignored all that and used idealized systems to explain the world. But that left most of the real world unexplained. The errors in Newton's calculations of planetary motion were ignored until Einstein came along to explain them. Traditional economics assumed perfectly rational agents. So does traditional survival training. Neither assumption reflects the messy real world.

The idea of chaos theory is that what appears to be a very complex, turbulent system (the weather, for example) can begin with simple components (water, air, earth), operating under a few simple rules (heat and gravity). One of the characteristics of such a system is that a small change in the initial conditions, often too small to measure, can lead to radically different behavior. Run the equations two, four, eight times, and they may seem to be giving similar results. But the harder you drive the system, the more iterations result and the more unpredictable it becomes.

Edward Lorenz, a meteorologist at MIT, was modeling weather systems on a computer in the early 1960s when he accidentally discovered that a tiny change in the initial state (1 part in 1,000) was enough to produce totally different weather patterns. That became known as the Butterfly Effect, "the notion that a butterfly stirring the air today in Peking can transform storm systems next month in New York," as Gleick wrote in *Chaos*.

Classical science aimed at predicting an outcome, then conducting an experiment to confirm it. But natural systems don't behave so neatly. The specific details can be described, yet no one can predict the outcome. You can describe how the weather works with high school math and physics, but you can't tell very far in advance when or even if it will rain. You can predict that lightning will strike under certain conditions, but you can't predict when or where. When I was a teenager, I teased my father by saying that from living with scientists all my life I had observed that they knew so much that they often seemed to know nothing at all. Classical science can't predict the behavior of a cloud, which is nothing but a bunch of water droplets moved around in the air by heat and gravity. Training and safety systems are also a form of prediction since they aim to control the future.

Chaos theory views such systems, which seem chaotic, as actually arising out of a simple, orderly set of mathematical functions. They may also produce effects that are the same at all scales. A cloud looks the same whether viewed up close or far away. So does a coastline. And much of what we call art appeals to the senses because of its so-called fractal nature. Nôtre Dame cathedral is beautiful at any scale. From far away, you can see the buttresses and the pleasing shape. The closer you get, the more interesting detail you see, until at last you are looking at the tiniest of figures sculpted on its surface. Matter itself appears to be like that. Just when you think you've found the smallest piece, you find another even smaller one.

The theory of self-organized criticality, sometimes called Complexity theory, was developed hard on the heels of chaos theory by some of the same people. It asked and suggested answers to questions as fundamental as: Where does order come from? How do you reconcile it with the second law of thermodynamics, which says that everything is heading toward more disorder? In a sense, complexity was an extension of the thinking that gave rise to chaos theory; indeed, it was often referred to as existing at "the edge of chaos." (There has also been strong objection to linking complexity and chaos and to using the term "complexity.") Like chaos theory, complexity theory postulated "upheaval and change and enormous consequences flowing from trivial-seeming events—and yet with a deep law hidden beneath." Complexity theory is a bold attempt to explain everything all at once, and so far it's done a better job in some ways than either Einstein's relativity theory or Niels Bohr's quantum mechanics did.

The climbers on Mount Hood discovered the enormous consequences of the trivial-seeming event of pulling their protection. Such upheavals are part of nature's mysterious tendency to create self-organizing systems at critical boundaries. The climbers did not recognize that they were part of a system that had reached that critical state where it would probably remain most of the time and where a seemingly insignificant force could set it going at any time.

PER BAK, a Danish physicist, set up an experiment in the 1980s that graphically demonstrates how accidents happen in wilderness recreation, though this was not his intention. He was demonstrating how a self-organizing system works. He created a pile of sand (or a computer model of one) and let more sand dribble on it from above as if from an hourglass. As the pile grew, it reached a certain height and then began to collapse. The pile didn't get any shorter, but it didn't get any higher, either. It simply continued in that steady state of continuous collapse. In his book *Complexity*, M. Mitchell Waldrop describes "the resulting sand pile as self-organized, in the sense that it reaches the steady state all by itself without anyone explicitly shaping it." No one had to design those collapses into the system. They were a characteristic of the simple system. It reached a state that some scientists call criticality (though, again, there is much argument about how to use this term, which has become jargon). "In fact," Waldrop adds, "the critical sand pile is very much like a critical mass of plutonium, in which the chain reaction is just barely on the verge of running away into a nuclear explosion—but doesn't." As more and more sand falls, there are many small slides. Now and then there is a great avalanche in which the mountain comes apart, but often nothing happens for a long time.

There is nothing at all in the physics of silicon dioxide that could predict the behavior of the sand pile. You could use physics and chemistry to examine a grain of sand until your lights go out and never discover the Sand Pile Effect. But the Sand Pile Effect expresses a tremendous amount about the way all of nature works. And it explains as well why Perrow came to regard accidents as a normal characteristic of certain systems. The systems he called complex and tightly coupled are actually self-organizing systems. The accidents are the collapses, if you will, in big technological sand piles, such as nuclear reactors and jet airliners. They all operate continuously in failure mode. Most failures—collapses—are small ones, such as a broken switch, a burned-out light, a busted rubber gasket, the glitches that we dismiss as normal. And they are normal. But like the temblors in an earthquake zone, they are also the quiet harbingers of the larger collapses that must eventually happen.

Small collapses are common on the sand pile. Large-scale ones are rare. But collapses of all sizes do happen with an inevitability that can be described mathematically as inversely proportional to some power of the size (with earthquakes it's the 3/2 power, which curiously is the same power as the one used to determine the time that planets take to go around the sun: the square root of the cube of the size of the orbit). Similarly, fender benders are common, while sixty-car fatal pileups are rare. But they both happen. Murder is common; six-state murder sprees are rare. Mountaineering falls are common; nine people falling into a crevasse with three fatalities is rare. That so-called power law is found extensively in nature. It's a more precise way of saying what Perrow was saying: Large accidents, while rare, are normal. Efforts to prevent them always fail.

Both the Sand Pile Effect and normal accident theory predict that space shuttle accidents in which the entire craft and crew are lost will happen, albeit with long intervals between them. The space shuttle *Columbia* disintegrated on approach to landing almost exactly seventeen years after *Challenger* exploded. Such accidents are inherent characteristics of that system. NASA will investigate and explain all the details of how it happened, but knowing those details will not prevent the next accident. Indeed, the safety precautions they take may make it more likely.

In connection with the *Columbia* accident, most engineers I spoke to speculated that the tiles on the underside of the craft, designed to absorb the heat of reentry, probably caused the problem. Dan Canin of Lockheed wrote in an e-mail, "Every precaution and material science known to man has been applied to the problem of making the thermal protection system work. It's a known risk. The tiles are soft, and every astronaut knows that if the wrong ones are damaged, the shuttle burns up. But the odds against it are pretty good, especially when compared to the rewards of being an astronaut, so they're willing to take the chance. In fact, they FIGHT for it . . . as would a lot of us. But getting the public to buy this is a lot tougher, especially a public that expects every risk in their lives to be mitigatable to zero. It will be interesting to see if NASA tries to take on this challenge, explaining to the public that doing bold things isn't about engineering risk to zero. Shit happens, and if we just want to restrict ourselves to things where shit can't happen . . . we're not going to do anything very interesting."

So the accident on Mount Hood was predictable; but no one could know which climbers would fall, where or when, or with what injuries. As with the sand pile, the overall system involving gravity, mass, and simple materials obeys rules that can be known.

Like the sand pile, the system the climbers used was poised at the edge of chaos, in a state of criticality. The small points of contact between each person and the mountain (crampons, the tip of an ice ax handle used as a cane) were like the interlocking grains of sand on the pile, set to release at the slightest touch. Every step was another chance for a slip—a collapse—of any size. Most were small—1 inch, 5 inches—and died out. At a less frequent rate, bigger slips occurred. Hillman saw Biggs fall that morning. He quickly arrested himself with his ice ax. There are ten thousand climbers on Mount Hood each year and only one death on average. The power law applies: The bigger the accident, the less likely it is.

I like Perrow's description of such accidents, because while he was talking about a nuclear power plant, he could just as easily have been talking about Mount Hood: "processes happen very fast and can't be turned off . . . recovery from the initial disturbance is not possible; it will spread quickly and irretrievably for at least some time. . . . What distinguishes these interactions is that they were not designed into the system by anybody." He's describing the self-organizing behavior seen everywhere in nature, which complexity theorists, such as Stephan Wolfram, believe probably gave rise to life and to us in the first place.

The climbers were familiar with the system and had a good concept of how it behaved, but only with its more frequent smaller collapses. Hillman's comment that he was ready to arrest five people falling 100 feet indicates how little he understood the amount of energy in the system. The large-scale collapse, when it came, did "happen very fast" and couldn't "be turned off." It did "spread quickly and irretrievably," and allowed them no chance to recover.

Less than forty-eight hours later, as Steve Kruse and I climbed up Mount Hood, we sat and rested on a rock between the lift and the crevasse. "This thing went critical faster than anything I've ever seen," he told me. Without realizing it, he was using the jargon used to describe self-organizing systems.

SEVEN

THE RULES
OF LIFE

THERE ARE REALLY two environments, two worlds, on Mount Hood. One is tailored for the comfort and survival of people. The other is not. There are the ski lifts, the lodge, and the five-star restaurant with its pinot noir and rosemary crostini. I'd seen it the night before at dinner. Timberline is a place where the chili is white and costs $6.75, the dessert is flourless, and the tea is Tazo. With my silverware clinking softly, I could possess those millions of square miles of wilderness with an unwary indifference that no other animal would dare. The mountain was safely contained within mullioned window frames. At a nearby table, two scientists, one from Los Alamos, the other from Livermore, discussed the tax laws of their respective institutions while sipping an amusing little fumé blanc. And there were the happy children on their snow-boards, riding the steel ski lifts, which lay like the bars of a cage to contain the beast.

But I can rule only my neat little model of that world. I saw how easy it was to cross that invisible dividing line between what has been adapted for us and what demands that we adapt to it. By the imperceptible increments of my footsteps, I changed the frame. The pretty lodge was reduced to a doll's house, while the wilderness, huge and voracious, exploded across my view. The rules changed, too.

I understood, then, how Bill Ward, Rick Read, and John Biggs, the three men who died, might have carried some of that attitude from the lodge to the mountain. Their success in life, their goals and plans and imagination, had brought them here. They'd made the money to do such things. They deserved the rewards that their mastery of life had brought them. Biggs was bent on climbing the highest peak in every state. So perhaps not even their combined experience could shake the view that Mount Hood was one more challenge to be met and managed.

Mount Hood would not be managed. I'd seen John Biggs's brother in the parking lot the day after the accident, searching John's truck for the ignition key so that he might take a few personal effects home to the family. He'd just come from identifying John's body—John, who'd survived flying a fighter plane in Vietnam and a career as an airline pilot, only to be killed by a "beginner's mountain."

The people were part of the mechanical system, but they were also a system in themselves. And without knowing any scientific theory, there was less esoteric knowledge that might have helped them that day.

All of the rescue professionals I spoke to at Mount Hood seemed to understand the fundamental mistake the climbers had made. Jim Tripp is head of the ski patrol for Mount Hood, and one day he invited me to his office to talk. I found it in the ski patrol room, an institutional-looking place with six bunks under orange, blue, yellow, and green bedcovers. There were stuffed animals on a high shelf and ski company stickers plastered all over the steel cabinets. (On the toe of one set of skis: "Danger! Air Bag Installed.") There were dispensers of Orthoglass synthetic splints, along with oxygen and IVs, cervical collars and backboards. This place was all about pain.

The only illumination in Tripp's cluttered office came from a

single desk lamp. Tripp is tall and lanky, and in this place of per-
petual snow, he wore bluejean shorts, sandals, and a gray sweat-
shirt. Feet up on the metal desk, cocked back in his secretary's
chair, hands folded behind his nearly bald head, he wore a big grin
as I came in. He looked like an aging surfer and knew as much
about Mount Hood as anyone. When I asked him about self-arrest
with an ice ax, he just laughed. "If there's only ten feet of slack, it
can pull you off," he said. "If you are a true master, then maybe,
maybe, you can self-arrest. The first rule is: The top man must not
fall. He's going to go one hundred feet before the rope is taut, and
you are screwed. This mountain is just not taken seriously. Fat people go up there."

Experience can help us or betray us. Bill Ward had had three to
five years of climbing experience and it led him to pull his protection, which led to his death.

Peter M. Leschak is a wildland firefighter and a gifted writer.
In *Ghosts of the Fireground*, he writes of waiting too long to get
out of a fire that was overtaking him. "I'd perhaps been too well
trained," he says. "I was a victim of a common fire service mind-
set: Can do! . . . A crew is ever anxious for the action and the nov-
elty. . . . Our instructor had drummed a phrase into our heads:
When in doubt, don't."

He's talking about a theory called "risk homeostasis." The theory says that people accept a given level of risk. While it's different
for each person, you tend to keep the risk you're willing to take at
about the same level. If you perceive conditions as less risky, you'll
take more risk. If conditions seem more risky, you'll take less risk.
The theory has been demonstrated again and again. When
antilock brakes were introduced, authorities expected the accident
rate to go down, but it went up. People perceived that driving was
safer with antilock brakes, so they drove more aggressively. With
the introduction of radar in commercial shipping, it was expected
that ships would collide less frequently. The opposite proved to be
true. Radar simply allowed the owners to require the captains to

drive the ships harder. Technological advances intended to improve safety may have the opposite effect.

Leschak said of his near-death experience, "We were normalizing risk: We'd been through a similar situation and had emerged just fine. . . . If you've tallied a lot of experience in dangerous, iffy environments without significant calamity, the mental path of least resistance is to assume it was your skill and savvy that told the tale." That same trap kills a lot of experienced climbers, skiers, hikers, boaters, cavers, and so on. Heraclitus said that every time you step into the river, it's a different river. Every time you walk on Mount Hood, it's a different mountain. To use the technical terms, it's a boundary condition, a phase transition zone. And because of that, even if you are intimately familiar with its subtleties of character, it can make a mockery of the most thoughtful plans. Experience is nothing more than the engine that drives adaptation, so it's always important to ask: Adaptation to what? You need to know if your particular experience has produced the sort of adaptation that will contribute to survival in the particular environment you choose. And when the environment changes, you have to be aware that your own experience might be inappropriate.

Climbers might read *Accidents in North American Mountaineering (ANAM)*, an annual summary edited by Jed Williamson. In his analysis of an accident that was almost identical to the one on Mount Hood, Williamson wrote that the three common factors in most mountaineering accidents are "(1) descending, (2) roped together, and (3) no fixed protection. A rope without fixed anchors invariably becomes the primary mechanism of multiple injuries during a fall."

Headlines in the *The Oregonian*, Portland's daily newspaper, would scream that it was just a "freak accident." It was not. It's a very common one, repeated over and over on numerous mountains.

The most experienced mountaineer on Hood put it this way: "A rope without fixed protection is a suicide pact."

. . .

THE PERSON is always a part of the system. Charles Perrow calls them "human-machine systems." In river running, you don't feel like being stationary. Once you're moving, it takes a lot of self-control to stop and scout or decide to portage, to disengage the human from the system when it may grow too chaotic. The same was true for the Canadian snowmobilers who were caught in an avalanche. Staying put was safe, but it didn't cut it emotionally. The same can be true of climbers descending.

There are three main difficulties with descent: attitude; an emotion involving goal seeking; and stress. In the first case, those climbers, like most, celebrated reaching the summit. "It was such a glorious morning," Slutter recalled. "We were enjoying the moment. We were joking around up there for half an hour." Laughter again: humor. The deescalating emotional response.

The trap lay in the fact that they were only halfway to their real goal. They were celebrating when they had the worst part of the climb ahead of them. Climbers are the only sportsmen who do that. Moreover, it is part of the natural cycle of human emotion to let down your guard once you feel you've reached a goal. When they took the Sno-Cat to the top of the Palmer ski lift, they were geared up for action, excited, and perhaps a bit anxious. The familiar adrenal response took shape to promote action. It powers the rising curve of energy until the goal is met (the hunt, the mating, the fight or flight). It was followed by a grueling climb of five hours. By the time they reached the summit, the chemicals of emotion had been metabolized, even as the body burned its reserves of glucose. They felt that the action was over. They didn't think it, they felt it. Laughter and celebration lowered their guard further. Energy took a downward curve toward rest and recuperation, just as it would following any other burst of activity. Their focus had been sharp in the goal-seeking phase. Now it grew blurry.

Various stresses contribute to falls on descent. Most climbers reach the summit tired, dehydrated, hypoxic, hypoglycemic, and sometimes hypothermic. Any one of those factors would be enough to erode mental and physical abilities. Put together, they make you clumsy, inattentive, and accident-prone. They impair judgment. They could, for example, contribute to a decision to descend without fixed protection. Tired muscles are less precise in their movement, making a slip more likely.

Stress can trump all the other effects. (The correct term is actually "strain," but "stress," mistakenly used by Hans Selye, a Canadian researcher, in the 1930s, has stuck as the accepted usage.) The stresses involved in wilderness recreation activities are more than enough to set about organizing an accident.

Last, there is the fact that descending is technically more difficult than ascending. During the climb up, your foot is planted before your body weight is shifted. The opposite is true on descent, and it's less stable. Descent, like the act of walking, is a controlled fall.

Even with unimpaired thinking, there's no guarantee that things would have gone any better. People routinely fail to realize that an accident not happening is no guarantee that it won't happen. As Scott Sagan puts it in *The Limits of Safety*, "things that have never happened before happen all the time." Unfortunately, as Perrow comments, "It is normal for us to die, but we only do it once." Which is too bad, for it might be the ultimate learning experience.

SO THE CLIMBERS faced downhill now with the Timberline Lodge in view. Suddenly, the high of celebrating the summit turned into the prospect of a long slog down to the lodge. *(The pinot noir, the rosemary crostini.)* Images of previous experiences blossomed across numerous areas of the brain. The climbers saw rest and safety close at hand: Just get down the spine to the lodge as quickly as possible.

Belaying is tedious, time-consuming, tiring. They were already tired, had already consumed enough time. Spinning rapidly and unconsciously through the options, they hit those emotional bookmarks, and one of those bookmarks may have reminded the climbers that the act of placing one foot in front of the other would soon yield comfort, safety, relief. Another bookmark told them that belaying involved prolonged pain, thirst, hunger, and fatigue.

The organism would not have been likely to vote for that. The climbers had no bookmark for falling down a 1,000-foot ridge, nor for the speed with which energy can build up in a system of ropes. Their successful practice at self-arrest would have worked against them, giving them an emotional certainty that it worked: They'd felt it work. The body knows. But they had felt the system only at low energy levels, where its behavior was predictable, not chaotic. None of them had ever tried to arrest something that weighed as much as a dump truck. None had tried arresting with a broken pelvis or a dislocated shoulder.

Unconsciously, they were asking themselves: How have I done this before? I roped up and walked and used my ice ax as a cane, and we got down all right. *(A hot shower.)* Okay, listen up: Just keep the slack out of the system . . .

Once the decision was made, the system accepted, those elements helped to form a new mental model. A basic assumption of the system was that someone could fall. So the idea that the system would hold a fall was embedded in the model. It was a given, an unconscious way of viewing the world. Then everyone got busy making sure that all the various parts of the system were connected and working properly. Tasks such as tying knots and checking harnesses would have occupied their attention, and that familiar and reassuring ritual would have substituted for a thorough analysis of the entire system or the model.

Slutter recalled: "We had just discussed with the group about keeping two points on the slope. We had reviewed all this the day before. Have contact with two crampons or a crampon and an ice ax at all times." Paradoxically, the very discussion that was intended to reduce risk encouraged faith in a faulty system.

So, piece by piece, unaware that their model of the world was no longer valid, they assembled the accident. And they began that process long before their arrival on Mount Hood. For even the accumulation of experience on other mountains betrayed them. As Slutter put it, "I have had to arrest myself more than once on many climbs and I know that you are going to move five to eight feet and then you are fine."

So they would have been getting the feel of taking precautions—ones with which they were familiar but which applied only at low energy levels. The model on which they operated, unlike the system itself, was stable: new information, such as the precarious condition of the ice, couldn't disturb it.

Slutter later recognized that fact. "I truly believe that the quality of the ice played a factor in us not being able to arrest," he told me after the accident. But he did not appreciate the forces he'd encountered. "I was in arrest position before there was any yank on my rope, and I was thinking: Don't worry, you're going to move five to eight feet. Then my ax went through the ice like it was a slush puppy. I went down the mountain on my chest. My chest was black from bruising. When I came to, I still had an ice ax in my hand."

Al Siebert, a psychologist, writes in *The Survivor Personality* that the survivor (a category including people who avoid accidents) "does not impose pre-existing patterns on new information, but rather allows new information to reshape [his mental models]. The person who has the best chance of handling a situation well is usually the one with the best . . . mental pictures or images of what is occurring outside of the body."

Everyone, to one degree or another, sees not the real world but the ever-changing state of the self in an ever-changing invention of the world. We live in a continuous reinterpretation of sensory input and memories, and they are contained in presets that can, at any given moment, light up neural networks in a shifting kaleido-

scope of energy, which we come to think of as reality. It is all part
of the dynamic dance of adaptation that accounts for our survival
as an organism and the survival of the species.

IN JUNE 1992, the National Outdoor Leadership School
(NOLS)—along with other like-minded organizations, including
Outward Bound and the American Alpine Club—organized the
annual Wilderness Risk Managers Conference, held at Conway,
Washington, to talk about accidents of just the sort that happened
on Mount Hood and to work out a methodology not only for
understanding what has happened in individual cases but for preventing accidents in the future. They discovered that wilderness
accidents follow a predictable pattern. Every accident investigator
I've ever talked to has expressed frustration at seeing the same
accidents recur again and again.

NOLS has attempted to codify common elements of accidents in
a matrix called "Potential Causes of Accidents in Outdoor Pursuits."
Dan Meyer, director of the North Carolina chapter of Outward
Bound, first published the Accident Matrix in 1979 in the *Journal
of Experiential Education*. Jed Williamson refined it. Contributing
causes of accidents are arranged in the Matrix under three general
categories: Conditions, Acts, and Judgments, which combine in a
dynamic and synergistic sequence to generate accidents.

It's easy to see how these categories would apply to the Mount
Hood accident. "Conditions" refers to any potential forces that can
hurt you, such as those resulting from a slip on a steep icy slope.
The main "Act" that set the sequence in motion was to pull the
protection and move while roped together. The "Judgment" was
the belief that the climbers could self-arrest, which in the Matrix
might be phrased as "overconfidence" or "exceeding their abilities." In other accidents, "Conditions" might be falling rocks, swift
and/or cold water, or weather.

That may seem self-evident, too obvious to be useful. After all, you don't need a matrix to tell you that if you're kayaking in a mountain stream, the water is going to be swift and cold. But such information can be a weak motivator. And as Peter Leschak, the firefighter, put it, "Sounds like a no-brainer, but in the heat of battle, simple concepts can wander off into the smoke and be forgotten." At the crucial moment for the climbers on Mount Hood, the information that belaying is safer could not overcome the urge to get off the mountain. Secondary emotions, emotional bookmarks, and mental models all conspired to encourage a sense of confidence about what the people were doing, even as stress worked to stifle any warning voice and mask cues about the changing environment.

A true survivor would be attuned to those subtle cues, the whisper of intuition, which might have been saying, *I don't feel quite safe here. Why is that so?* But since most of us are not conscious of those processes, we have nothing to draw our attention to what's happening to us. We don't have what psychologists call metaknowledge: the ability to assess the quality of our own knowledge. It's easy to assume that perception and reason faithfully render reality. But as Plato suggested and modern neuroscience has proved, we live in a sort of dreamworld, which only imperfectly matches reality. None of the mountaineers on Mount Hood said to himself or his companions: *Oh, my God! I'm being told to pull my protection by a mysterious force! Help! Help!* They literally never knew what hit them.

Some victims still don't know. One man who had decided to do some South American river rafting and got lost in the Bolivian jungle for three weeks during flood season wearing nothing but jeans, a T-shirt, and street shoes told me that, "apart from a personal machete, we were well equipped." After the Mount Hood disaster, which was widely reported, dozens of people were once again heading up (or down) the Hogsback in roped teams, one above the next, without protection. As you read this, the same acci-

dent has already happened again in more or less the same conditions on Mount Rainier. The helicopter crashed, too.

This isn't an argument for using belays or for not climbing. It's
simply an argument for knowing. To rope up is a serious decision.
I've had to make such decisions all my life in pursuits ranging
from riding motorcycles across the Mexican desert to flying bush
planes in the Arctic. But there is a subculture among mountaineers, and for most, the rope is more than a safety device. It's
both a real and a symbolic commitment to a partner. Hillman,
whose partner Biggs was killed, raised the rhetorical question:
Given the chance, would he have considered cutting his rope? For
him, the answer was clear. "No. There was no question what I had
to do as a member of a team and as a friend."

Nevertheless, it's worth considering the next Matrix category,
"Unsafe Acts." One item on Williamson's list is "Unauthorized/
Improper Procedure." Again, that may seem obvious now, but if
you wait until you're tired and hungry and thirsty, when all you
can think of is a hot shower and a cold beer, then you may be
tempted to just *get there now*.

Innumerable accident reports contain sentences such as "The
pair decided to solo the first few 5.5 [i.e., easy] pitches to save time"
("solo" means they climbed without placing protection); or "The
group was in a hurry to descend after spending longer on the ascent
than they had expected." Or a classic: "He took a shortcut . . ."

Under the third column in the Matrix, "Judgment," numerous
examples seek to explain the decisions that led to accidents: peer
pressure, schedule, misperception, disregarding instincts . . . And
they all fit, too.

There may have been peer pressure in the Mount Hood accident. If any one of the climbers had serious doubts, if anyone
heard that small voice, no one said anything. In such situations,
group dynamics can be a powerful motivator. It is well documented that co-pilots aren't likely to challenge pilots in aircraft

cockpits and sailors aren't likely to challenge captains, sometimes with fatal consequences. Experienced climbers may be reluctant to challenge others with experience, and the same is true in any other pursuit. Going into a risky operation, doctors won't challenge doctors. Going into a risky situation, cops won't challenge cops. So it's no surprise that none of the climbers wanted to be the one to say, "I can't handle this," and force everyone to belay and delay. You can just imagine back at the bar, all of them laughing, saying, "Yeah, Rick here made us belay all the way down the frickin' Hogsback. It took *forever*."

Even so, I tend to agree with Perrow: "I am arguing that constructing an expected world . . . challenges the easy explanations such as stupidity, inattention, risk taking, and inexperience." Perrow is proposing the frightening idea that we are doing the best we can with what we've got. We are not asleep at the switch. We are doing what everyone does, what the best among us do, and when we have such an accident, it's normal.

The design of the human condition makes it easy for us to conceal the obvious from ourselves, especially under strain and pressure. The Bhopal disaster in India, the space shuttle *Challenger* explosion, the Chernobyl nuclear meltdown, and countless airliner crashes, all happened in part while people were denying the clear warnings before them, trying to land the model instead of the plane. The Matrix is an effort to manage wilderness risks using some of the same analytical tools that have been used in industry for more than a hundred years.

Carl von Clausewitz introduced a related concept in *On War*, published posthumously in 1833. His text has been a classic guide for generals ever since and is still taught at military schools around the world. Clausewitz writes of "countless minor events" that "conspire to decrease efficiency, and one always falls short of the goal. These difficulties happen over and over again, and cause a sort of friction." He's talking about an army in the field, which is not altogether different from groups of people in the wilderness, where the qualities Clausewitz identified as ideal in a general can come in handy.

Clausewitz observed, "The military machine . . . is basically very simple, which makes it seem easy to manage." Again, some simple systems are capable of complex behavior.

> But we must remember that no part of it consists of a single piece, that everything is made up of individuals, each of whom still has his own friction at every turn. . . . The battalion is always made up of a number of men, the least significant of whom may very well bring things to a halt or cause things to go awry. . . . Therefore, this terrible friction . . . is everywhere in contact with chance, with consequences that are impossible to calculate, for the very reason that they are largely elements of chance.

And most important for those undertaking such challenges as nature presents, the army general "must have knowledge of friction in order to overcome it, where possible, and in order not to expect a level of precision in his operation that simply cannot be achieved owing to this very friction."

When my eldest daughter, Elena, was about six years old, she and I decided that we were going to write a work we'd call *The Rules of Life*. The first rule we came up with was: *Be here now*. It's a good survival rule. It means to pay attention and keep an up-to-date mental model. The second rule was: *Everything takes eight times as long as it's supposed to*. That was the friction rule, which travelers in the wilderness will do well to heed.

There is a tendency to make a plan and then to worship the plan, that "memory of the possible future." But there is also a tendency to think that simply by putting forth more and more effort, we can overcome friction. In the case of the Mount Hood accident, both happened. With the large number of people on the mountain, it became like a battalion. Communication and movement were difficult. Separate teams couldn't discuss and modify their plans as conditions changed. Each person had his own friction, each team had its own friction, and the cumulative effect put everyone on the same slope roped up at the same time with no protection and with the ice melting.

Rather than accept friction as a fact of life, they tried to overcome it. And as history shows, the harder we try, the more complex our plan for reducing friction, the worse things get.

Because the larger system that is made up of the vast population of people now involved in wilderness recreation is subject to normal accidents, the overall rate of accidents is likely to stay the same. But individuals can take steps to avoid them. To admit reality and work with it is to accept it. *Be here now.* And plan for everything to take eight times as long as you expect it to take. That allows for adaptation to real conditions and survival at the boundary of life and death, where we seek our thrills.

The seemingly simplest accident is complex, if only because it involves the human brain. The simplest physical systems may be capable of staggeringly complex behavior, too, as was the case on Mount Hood or in any river-running accident. The interactions of a river and the boaters on it challenge even the tools of higher math. When such nonlinear physical systems meet emotional systems, the results can be gruesome.

In the summer of 1997, Ken Phinney was paddling on the Chattooga River on the Georgia–South Carolina border. Although he wasn't the most experienced river runner, he was with more experienced people, who tried to help him. The accident report stated:

> Arriving at Crack in the Rock Rapid, the lead boater ran Right Crack and set up safety. Ken and his friend, despite strong warnings from the lead boater, beached their boats on river right and swam across the river so they could scout Left Crack. This drop is filled with undercut rocks and has killed several others in past years. Phinney didn't make it across, and washed into Left Crack

feet first. He was pinned horribly under several feet of water. It took rescue squads several days to pull him out, but his body was torn apart in the process. . . . Left Crack is a well-known danger spot, and the victim was warned.

Simple forces power this complex environment. We cannot escape them, for they have formed us and everything in our world.

The climbers on Mount Hood were set up for disaster not by their inexperience but by their experience. It was the quality of their thinking, the idea that they knew, coupled with hidden characteristics of a system they had so often used. The system, like the sand pile, was capable of displaying one type of behavior for a long time and then suddenly changing its behavior completely. All it took was what Gleick calls "a kick from outside." That kick came from Bill Ward's slip, but it could have come from anywhere. The only certainty was that it would come, somewhere, sometime. But while such large-scale collapses are inevitable, the involvement of those particular climbers in the event was not. Any one of them could have disassembled the system at any point before Ward slipped simply by untying a knot or putting in protection. Such accidents have to happen. But they don't have to happen to you and me.

EIGHT

DANGER ZONES

ONE SUNNY DAY, I boarded a United Airlines flight in Chicago and flew to the island of Kaua'i in the state of Hawaii. A pretty woman put a wreath of orchids around my neck, I grabbed a ride, checked into the Holiday Inn, and got into my swimming trunks. Then I went down to Lydgate Beach, which was right outside my window. I wanted to wash off the airliner stink while body-surfing those big breakers I'd seen on final approach. As I hit the sand in my flip-flops, I could see great ranks of creamy waves curling in toward shore from the deep blue water. I couldn't wait. I passed below Tower Number 5, heading toward the surf, and the life-guard on duty there said hi. I thought: People sure are friendly here. I said hi and asked him how the swimming was.

"Well," he said, as if considering it for the first time, looking out to sea, rubbing his goatee. He was a redhead, thin and freckled, his face lined with experience. He came down from the tower and studied the sea. Before answering, he seemed to think of something and asked my name. I told him. He grabbed my hand in a surfer's handshake and said, "Mike Crowder." I waited. He turned back to the sea and continued to stare at it with a spaced-out look, and I started to wonder if he was stoned. Hawaiian weed is supposed to be bitchin'.

"Okay," he said at last, as if he'd decided something. It was only then that I realized he'd been reading the waves, the lineup, the break. "See where that flat water is over there? I'd say you can go in there and be okay. But you get more than about ten yards off shore, the rip will carry you out . . ." His arm was out, moving left ". . . there," and he pointed beyond a ring of rocks, waves crashing against them, which had been piled up to form a protected area for (I assumed) little swimmers, like a play area. There were families in there, laughing and tossing the kids around. Crowder was still pointing: "Then the current will carry you around that swimming area." All this very matter-of-fact, as if giving me driving directions. "It'll bring you back in over here, and just beat you to death against those rocks. So I'd say either stay in this protected area or else don't go more than ten yards off shore. Don't go in any white water at all. And you should be okay if no big waves come in."

And I thought: Wow, this guy really knows what he's talking about. And: If he hadn't been standing there, I'd have jumped in and swum right out into those big, happy breakers. It all looks so . . . inviting. And I began thinking how easy it is to die. Never to be seen again. I left my safe home, went to the airport, and there I was, utterly clueless, heading for the bottom of the food chain. Mere chance in running into Crowder had saved me.

A few days later, I was snorkeling. I had kayaked with a guide to a secluded beach, and she told me the same thing: The current just a short distance out would carry me beyond our sandy beach, and then the cliffs would prevent me from getting back to land. The next put-in was six miles. "You'll be all right if you can tread water for six miles," she said. And I started to think: Whoa. Paradise is some serious business. Or as the *Tao Te Ching* puts it:

Heaven and earth are inhumane;
   they view the myriad creatures as straw dogs.

A scientist named Chuck Blay did extensive research on death by drowning in Hawaii. He found that the efforts of people such as Mike Crowder had little to do with the rate at which people drown. Instead, the rate was directly proportional to the number of tourists who came to the island. Put another way, drowning is normal. Of the many drowning victims he'd studied, 75 percent were visitors and 90 percent were white males in their forties or fifties. Dr. Blay said, "The profile is: A guy comes here. He looks out there. And it's so beautiful and blue and inviting that he goes in." In other words, me. "The water's warm and pleasant, and it seems benign. But what he doesn't think about is that we're in the middle of the ocean here, two thousand miles from the nearest land. There's nothing to stop the waves or slow their momentum." Just off the reef, he said, the sea is four miles deep. A wave, like a ship, can have most of its mass below the waterline. "These waves come in from the deep open ocean," Dr. Blay said, "and there's nothing to stop them until they reach those cliffs."

As he described the scenario, I could picture myself a bit farther out than I'd intended to go, right in front of Crowder in Tower Number 5, trying to swim back in. I can't make headway against the current. Now my amygdala gets into the act, and I get all excited, thrashing around. I start to wear myself out. I swallow some water. It's salty and warm and makes me sick. Gasping, vomiting, I swallow even more. And it makes me feel heavy, very heavy.

I'm still alive, but I'm being carried around the island now. If I try to get out on the cliffs, the waves slam me against them. The rocks are sharp lava, and the waves are large. A single wave can kill a swimmer trying to climb out. Most people are never found because the tiger sharks that prowl those waters like to chow down after the wave action on that volcanic glass turns swimmers into human chum.

Dr. Blay told me about a certain rock where tourists liked to photograph each other because the crashing waves look cool. "You go to adjust your camera," he said, "and when you look up, your wife's gone. And she's *never seen again.*"

One of the things that kills us in the wilderness, in nature, is that we just don't understand the forces we engage. We don't understand the energy because we no longer have to live with it. Like Huskisson being run over by the *Rocket*, we fail to grasp speed and distance. Even when we're told, even if we understand it at an intellectual level, most of us don't embrace the facts in that emotional way that controls behavior.

The environment we're used to is designed to sustain us. We live like fish in an aquarium. Food comes mysteriously down, oxygen bubbles up. We are the domestic pets of a human zoo we call civilization. Then we go into nature, where we are least among equals with all other creatures. There we are put to the test. Most of us sleep through the test. We get in and out and never know what might have been demanded. Such an experience can make us even more vulnerable, for we come away with the illusion of growing hardy, salty, knowledgeable: Been there, done that. In *Into Thin Air*, Jon Krakauer wrote about a guide, Scott Fischer, who had encouraged him to climb Mount Everest. "We've got the Big E figured out," he told Krakauer. "We've got it totally wired." Fischer died up there. One of the great Stoic philosophers of the later Roman period and Marcus Aurelius' teacher, Epictetus wrote, "Be silent; for there is great danger that you will immediately vomit up what you have not digested."

Paul Fussell, in his book *Wartime*, describes the stages of enlightenment that beset a soldier as he progresses from green recruit to combat veteran. First he thinks, "It *can't* happen to me." Then he sees action and it becomes, "It *can* happen to me, and I'd better be more careful." And finally he sees enough of his fellows die to realize, "It is *going* to happen to me, and only my not being there is going to prevent it." But in wilderness recreation, where we are usually not in such close contact with death, our experience is subject to looser interpretation.

The psychology of oblivion is nothing new. The classics scholar Carolyn Dewald, writing of Herodotus, notes how the characters in the histories "come to grief because they have not paid attention to facts about the world that would ultimately defeat their plans. . . . Sometimes they commit errors that simple enquiry would have avoided. . . . Cambyses sends his army off on a mad march across the desert against the Ethiopians without sufficient information about the terrain."

TWO DAYS before I arrived in Hawaii, two twenty-one-year-old girls—tourists—went out for a day hike. It was the middle of August and the weather was mild and clear in Sacred Falls State Park on Oahu, just south of Kaua'i. They parked their bikes in the shopping center. They carried three water bottles and some snacks, intending to be back before dark. When I left the island ten days later, their parents had flown in to keep a vigil. The girls still hadn't been found.

During 1994 and 1995, three hikers vanished in the same area and were never seen again. During one of the rescue efforts, which involved 150 people, three searchers were lost as well. People just disappear sometimes. Those stories cry out for more information. We want to know what happened, but we can't. They remind me of something Michael Herr wrote in *Dispatches*, his classic book about Vietnam. A long-range reconnaissance patroller told him a combat story that went like this: "Patrol went up the mountain. One man came back. He died before he could tell us what happened." Herr wrote: "When I asked him what had happened he just looked like he felt sorry for me, fucked if he'd waste time telling stories to anyone dumb as I was."

In part, we miscalculate the scale of the places we elect to explore. And distance means energy, the energy we'll run out of if we're not found; the energy it takes to get out. Like Cambyses, we head off on "a mad march . . . without sufficient information about the terrain." But information is one thing. Believing it is another. There's nothing in our everyday experience to compare with the potential powers we confront in our quest for fun. And if

our experience happens by chance to be benign, then, as George
Orwell wrote in *1984*, it leaves "no residue behind, just as a grain
of corn will pass undigested through the body of a bird."

WHEN I was growing up, my family spent summers on Galveston
beach. My brothers and I body-surfed breakers there that looked
much like the ones on Kaua'i. I learned in my earliest years how
benign they were, how good it felt. But the Gulf of Mexico is not
Kaua'i. So, while I knew the rewards of Galveston, I didn't understand the hazards of Kaua'i. The act of walking into the forest,
swimming in the surf, or floating on a river does not seem threatening. Putting one foot in front of the other, looking around at the
orchids growing and the birds crying, it would never enter our
mind that we might *never be seen again.* It's . . . unthinkable.

Chuck Blay described to me what might have happened to
some of those lost hikers. The cliffs on the islands can be as high
as 3,000 feet. In spots around the circumference there are towns.
But the interiors of the islands are wilderness, triple-canopy jungle, vast and voracious—uncharted volcanic landscapes gone
mad with a chaotic overgrowth of flora and fauna. Once you
enter there, you can't see the sun. There's no sense of direction.
Paths peter out to swamp and vine thickets. Great carpets of rot-
ting vegetation grow out over the cliffs, cornices of plant life,
thrusting cantilevered into empty space. It's not uncommon for
people who get turned around in there to hear the surf below
and walk toward it, thinking of escape. They venture out onto
that spongy shelf, believing that they're still on solid ground.
They happen onto a soft spot, punch through, and fall 4,000 feet
to the sea. They die on impact and are then churned against the
volcanic broken glass of the rocks. They are eaten by sea life and
are never seen again. It's difficult to think of such things as you
leave your bike in a mall parking lot on a sunny day on the vacation of a lifetime.

. . .

WHEN I was learning rock climbing, my instructor warned me about trusting the stability of rock, any rock, even the beautiful crystal-studded granite that we were climbing in the Needles in the Black Hills of South Dakota. She told me of a couple, friends of hers, who were rappeling off a big slab of granite on the Matterhorn. It was an enormous boulder. That same rappeling position had been used for generations, she said. A sling was thrown around the slab for an anchor. Moments before, the couple had let two other climbers rappel ahead of them because they seemed to be in a hurry. Then the wife let her husband go first. As she watched him descend, the great slab began to slide. Instinctively, she reached for him. Their fingers touched. The slab seemed to float in the air, as if it had no substance at all. It shifted, then glided over the edge, blotting out the view of her husband, making her a widow, and changing forever her perception of the solidity of mountains and the orbital pull of risk and reward.

For most people it's unthinkable to imagine what appears to be a solid mountain coming apart. But all mountains are in a state of continuous collapse. The disconnect between that reality and our perception leads to many accidents. Being hit by rockfall or having a rock come loose in your hand is common, because people don't believe that the mountain is falling. Yet opening your eyes proves it: Mountains have skirts of scree and boulders that show it. In that deepest place of belief, it's easy to persist in thinking that the mountain is solid. For the most part, experience reinforces that mistaken impression.

The exiguous nature of everyday experience creates habits of mind that shape perceptions, so you don't see the mountain crumbling. Human life is so brief in comparison to the mountain's. You live in a compressed time frame that does not match many phenomena you encounter in nature. It takes a concentrated effort to see the sun crossing the sky, which it does in half a day. Yet your perceptions see it as hanging stationary in the sky. Only by forgetting to watch it do you notice, after the fact, that it has moved. So with the mountain.

Making someone into a believer, grasping the forces of nature, is difficult. Sometimes it takes a near-death experience. Some people are so resistant to conversion that even after the near-death experience, they remain unconvinced. Few of us believe in our own mortality until we're face to face with it, and then some of us forget immediately afterward. So we have no way to prepare for what seems too remote a possibility. And as Christopher Burney, who was a prisoner of war at Buchenwald, said, "Death is a word which presents no real target to the mind's eye." The best way to become a believer, short of dying, is to sit very quietly and contemplate those things.

QUIET IS best because some people don't catch on even after repeated experiences. They are too caught up in the noisy action to hear the voices. Butch Farabee, National Emergency Services Coordinator for the National Park Service, reported, "In Yosemite one summer we rescued a guy with sunstroke from a big wall. The Navy came in with its helicopter and we pulled the guy off with a 400-foot rope. Six months later, on the other side of Yosemite valley, we pulled the same guy off a mountain. This time he had hypothermia, which killed him. From one end of the valley to the other, from one temperature extreme to the other, this guy underestimated the conditions." Lots of people like near-death experiences. It's a thrill that can become addictive. But most don't want to get quite that near.

When you cross a busy street, you make calculations involving speed, distance, time, and so on, but you can't explain how you do it. When driving a car, you make about four hundred observations for every two miles you travel. You have to make about forty decisions. And you also make one mistake. You aren't born knowing how to drive, and you aren't naturally able to cross a busy street. Children can't. They have to be taught. But our culture places such great importance on that type of learning that, by the time you grow up, you don't even have to think about crossing the street. People from other cultures, say Polynesia, might not be able to do that; but they, unlike us, would be able to navigate the open ocean in a raft and hit a tiny island spot on. That kind of learning is important in their culture.

Going into the wilderness is, for many of us, like the invention of the steam locomotive was for poor Huskisson. The unheard-of energy level produced by the new invention was beyond his well-trained imagination. Most of us in the outdoors don't understand the energy levels in nature any better. Put us on a raft among the Polynesian Islands, and most of us would die. But people routinely go to places and engage in activities every bit as hazardous (in some cases more so) with little or no training, plan, forethought, or equipment.

I know the Vermilion River in my home state of Illinois. It's not exactly the Grand Canyon. But in mid-June 1996, a rainstorm had raised the water level, and on the 15th, six rafts were shooting down it at a good clip when they hit a dam. It wasn't even much of a dam, just 5 feet high, and there was a fish ladder on the right, an angled portion, that let five of the six rafts get down safely, despite their miscalculation about the changing conditions. But one raft was caught out of position and flipped into the hydraulic at the bottom, an area of stable turbulence common to most dams. The report stated: "One man, who was wearing a PFD [personal flotation device], washed out. The other man, James Schiro, was not wearing his PFD. He disappeared. His brother obtained permission to drop dynamite into the hydraulic to release the body, but this was not successful. The body was found three days later, many miles downstream." As for the notion that, with all those people around, someone might have helped Schiro out, Charlie Walbridge, of the *River Safety Report*, warns, "Dams kill almost as many rescuers as they do paddlers!"

There are places like that everywhere—places you can get to far more easily than I reached Lydgate Beach. (The Vermilion is about a two-hour drive from my home.) Without technical climbing abilities or gear, for example, you can reach the top of Mount Hood, Longs Peak in Rocky Mountain National Park in Colorado, Mount Washington in New Hampshire, and even Denali. All you have to do is put one foot in front of the other. Two hikers were on Longs Peak in February 2000 when a 150-mile-an-hour gust of wind picked one of them up and threw him down on the rocks. He weighed 170 pounds and was wearing a 70-pound pack on his back. He was banged up but survived. Gusts of 220 miles an hour have been clocked up there on two different occasions. And that's not the windiest place in the United States.

Lightning also claims many victims on Longs Peak. Like Cathedral Peak in Yosemite, it produces intense orographic lifting on summer afternoons. On July 12, 2000, a bolt of lightning went right through Andy Haberkorn's chest while he was climbing there and killed him. Despite the old saying to the contrary, lightning does strike twice in the same place. Jim Detterline, a Rocky Mountain National Park Ranger, reports: "Lightning is common [in] the Colorado Rockies high country. Haberkorn's fatal strike occurred about 15:00, a most dangerous time of day to be high on the face in midsummer. . . . There is a past history of a climber struck by lightning at this same spot."

The forces we engage are relentless. Gravity is on duty all the time. And nature's many less obvious forces continuously arrange systems capable of releasing tremendous energy, from storms that appear suddenly in high terrain to big waves on pretty beaches. Similarly, it's difficult to comprehend the energy that is entrained in snow moving downhill. We find it difficult to imagine the constant, urgent need for oxygen and water to run those electrochemical reactions that keep us alive. Although there's wide variation, the rule of thumb is that you can survive three minutes without air, three days without water, and three weeks without food.

When you leave the low-risk environment of home and place yourself in one where the forces are beyond ordinary experience, you engage systems that kick back when you least expect it. Such kickback systems, as I've come to think of them, like the sand pile, can be upset (or set up) by very small inputs. It's easy to underestimate the energy and the spaces out there. Ski resorts, for example, create the illusion of safety in the midst of wilderness. People approach them as if they were amusement parks, with little idea of where they are and what forces they may encounter.

NICK WILLIAMS went up for a few hours of skiing at Squaw Valley in California on December 19, 1998. You don't exactly have to prepare for a polar expedition on a Saturday morning ski run, do you?

Or do you?

Williams, who was fifty-one at the time, told me, "I was skiing by myself. The weather was forecast to be in the thirties, with light snow. I didn't have on very good ski gear because I thought I was out for just a couple of hours in mild weather." He was wearing a Dallas Cowboys jacket and not much else. "It was the first time I had skied at Squaw."

He'd gone up the Granite Chief chair lift on his fourth run when a blizzard came up. He found himself in a whiteout. As I had done once before, Williams just wanted to get back to the lodge. He was ignoring the Rules of Life. He wanted to *get there now*. He checked his crude resort map of the ski runs and decided "I could cut through some trees onto an intermediate run to get to the lodge."

Williams started down and found himself at a 500-foot cliff with a valley below. He was no longer at a ski resort. He was stranded in big-time wilderness without clothes, matches, water, or food. He had nothing.

All afternoon, Williams worked his way up hills and skied down them, but he couldn't reorient himself. "There was not much more I could do on skis," so he left them, kept his poles, and began hiking. As night fell, he propped himself against some trees. "Whenever I'd fall asleep, I'd fall away from the tree," he said, which kept him awake. He knew if he fell asleep, he'd die. He melted snow in his mouth for water. Periodically, he'd get up and do calisthenics to stay warm. It was bad. But he was doing something about it.

The blizzard continued into the next day; Williams was slogging through snow up to his mid-thighs. He inadvertently crossed a snow-covered stream, and like the man in Jack London's story, he broke through the surface and filled his boots with water. Then his feet froze solid. He tried to build a fire. "An hour before dark that evening, I gathered kindling and stacked it against a granite rock face." Williams then crumpled up the money he had in his pocket and struck his ski pole tip against the rock for sparks, but he couldn't start a fire. He struck it until it broke. He later learned that if he'd been able to make a fire and thaw his feet, they would have had to be amputated.

His second night, the thermometer at Truckee hit 19 degrees below zero, which meant that where Williams found himself, it probably reached 50 below. By all rights, with frozen feet, he should have died. As it was, Williams went into serious hypothermia, shaking uncontrollably, thinking of his family, trying to muster his will to live, and praying: "Dear God, take my feet, don't worry about that, help me get through this thing." He had a son back home, and he wanted to see him again.

His wife and son were in Florida on Christmas vacation, where Williams was supposed to meet them a week later, so they didn't even realize he was missing for a time. When he missed appointments and didn't check his mail, people knew something was wrong and contacted his wife. She was told by authorities that no one could survive a night out in such cold dressed as Williams was and that she had better make funeral arrangements and get her finances in order.

At first light on the third day, the blizzard stopped, the sun came out, and Williams doggedly started out again. The sun topped the ridge at 11 A.M., and he leaned against a tree and let it warm him.

An hour later, he heard the snowmobiles, and a little while later he was short-hauled off by helicopter. "When my son heard my voice," Williams told me, "he just started crying, and so did I." Williams still has three toes left. He also has a rare and precious knowledge of the world.

Many factors contributed to Williams's surviving what would have killed most people. He was extremely fit, jogging and roller blading the steep hills of San Francisco. He was a graduate of the Naval Academy and had been a Marine fighter pilot. And he'd had the drive, strength, and character to climb the corporate ladder; he was president and chief executive officer of Premisys Communications, a company with $100 million in annual sales at the time. So, as a spokesman for the hospital that treated Williams put it, "His mental toughness was unreal."

All of that contributed to Williams's survival. But I think the key thing was his son. I believe that Nick Jr. brought his father back, though he might never know it unless his own son does the same for him one day. I couldn't help thinking about Nick Jr. hearing Williams's story over and over and then wondering why his father survived an environment that would have killed most men. Would he believe that his father was more than human? Would he go looking for that right stuff, that hard-soft, willful-flexible adaptability that had kept his father alive?

KNOWLEDGE OF the sort you need does not begin with information, it begins with experience and perception. But there is a dark and twisty road from experience and perception to correct action. Unable to understand the forces they engage, unaware of their own position and condition, people can blunder blindly into harm's way. They see places like Squaw Valley, Aspen, and the Potomac River as amusement parks and even find themselves approaching vast wilderness areas with the same frame of mind. And often the hazards out there are begging for attention. The places we go are not named at random. The names may mean something. Forbidden Wall is among "the hardest and least repeated" in Zion National Park.

Jed Williamson told me, "There is a place called Skillet Glacier, on Mount Moran in the Grand Tetons, which had a repeating pattern of incidents." The hazard is well known: Skillet, get it? People climb up in the morning and by the time they're ready to descend, the snow has softened and become wet. "They start to glissade and lose control. I call this the Wile E. Coyote Factor," Williamson said. "You know, the last thing he does is to give the audience a big, self-satisfied grin. Then he looks down." There are numerous places whose names suggest their sad histories.

From below, Lambs Slide, on the east side of Longs Peak in Rocky Mountain National Park, doesn't appear that treacherous. And it's not, until you're fully committed about one third of the way up. Even in mid-July, when it's at its worst, the initial ascent is a deceptively easy hike. But it gets steeper and steeper. As on Mount Hood, people go up roped together, because it feels safer. (A team leader from Portland Mountain Rescue told me, "All that rope means is that you won't die alone.") The accident that results is so common it would have an acronym, only it's not pronounceable: LFCSAFTD. The official phrasing in the reports is usually "Lost Footing. Couldn't Self-Arrest. Fell to Death."

Williamson wrote: "Since the epic 1871 uncontrolled descent of Lambs Slide by the Reverend Elkannah Lamb, the accident has been repeated continually with results varying greatly from no injury to death." On July 14, 1996, Nathan Dick slid 1,000 feet and stuck his ice ax through his neck. Despite the existence of four other much easier descent routes, seventy-five people have died at Lambs Slide since Elkannah Lamb gave it its name. The death rate isn't that high, but it nevertheless means that a lot of people have been unpleasantly surprised in a pretty place that has a reputation as a beginner's peak. And the frequency of fatalities probably follows a power law.

We are human. Our attention is fragmentary. We get excited. We get tired. We get stupid. Of course, you can't make adventure safe, for then it's not adventure. In an almost comic treatment of the paradox, the Mount Hood recreation officer told me, "If you made it so safe for everybody to get up there, you'd have a lot more fatalities because people wouldn't recognize the risk." More likely, people wouldn't bother to go up. But we want to go up. We want to conquer. I don't mind dying. I just don't want to do it today. And I'd hate like hell to have my gravestone read: "Here lies a moron."

I HAD been on Kaua'i a few days, and I'd started taking surfing lessons. My instructor, Ambrose Curry, a local legend, told me, "If you're not afraid, then you don't appreciate the situation." One day, Mike Crowder and I were talking on Lydgate Beach, and he told me, "You get to the point that when you see your first set in the morning, you know exactly what it's going to be like. Then you get in the water and you can feel it and you know exactly what to do. It's a very spiritual thing." He told me that *mana* was the Hawaiian word for "spirit" or "spiritual power" or "energy." If there needs to be a reason for surfing, *mana* is the reason, he said. "Surfing is about managing energy." He described the ocean as a pure analog, a graph, of energy patterns. And I began to realize how sophisticated his knowledge had become, this part-time lifeguard who made a living doing odd jobs while his wife sewed baby clothes. Who made do in order to surf.

"Sometimes I see a surfer out there," Mike said, "and he's got his timing just right, coming down the face of a wave, and when he comes back in, he walks by me on the beach, and I can just feel the *mana* coming off him in waves as he goes past. It's spooky. Surfing is a very spiritual thing." Crowder had discovered an ancient cultural survival technique in the sport of surfing, a way to learn—physically—about those forces that might kill him. Through physically interacting with the energy patterns, he had come up with a way to detour around some of the brain's road-blocks and blind alleys and get the knowledge into his bones and nerves and muscle fibers, where it could stand guard for him. He'd matched his body's constant drive toward homeostasis, his adaptations, to his environment. Even the experts get caught, though.

Surfers remember their big wipeouts vividly, like God handing them tablets on the mountain. (They remember those of other surfers, too: Komo Hollinger's near-death wipeout at Waimea in 1969; Owl Chapman's 1970 wipeout, which was so brutal that readers of *Surfer* magazine wrote to complain about it being featured on the cover. It made surfing look, well . . . dangerous.) As Mike and I were talking, he showed me the scars in his scalp and on his back and legs, hieroglyphic transcripts of his apprenticeship. The sea gods had tattooed his diplomas right on his skin. He had evidence of the forces he'd encountered on the outside of his body, where he could see it, literal emotional bookmarks, if you will.

One day on a big wave, Mike said, he "got a bad feeling," but wasn't able to act fast enough. It was a subliminal perception, an emotional bookmark that he wasn't able to jerk into conscious action before he was slammed into a sandbar so hard it nearly broke his neck. If it had been reef instead, he'd be pure *mana* now. He surfs without a leash. "You lose your board, you swim," he says. And that could be a half mile in 18-foot seas.

At the larger beaches, the lifeguards use Jet Skis, without which they'd never reach a victim in time. At Tower Number 5, which is kind of a run-down neighborhood where people such as writers go, they have no Jet Skis. Mike has to swim if he's going to rescue someone.

Speaking in a slow, serious, measured fashion with lots of pauses and emphasis, Mike mused: "Sometimes . . . I'm *life* guarding and I *see* somebody out there . . . and I just *know* they're going to do something . . . *dangerous*. I do a lot of praying for these people, and it usually works . . . I *pray* them back in. . . . So I do more . . . *prevention* than actual life-saving."

One morning, Mike's wife Limor (which Mike tells me means "My fragrant one") brought their two curly-headed toddlers down to the beach to watch their father surf at 7:00 A.M. When Mike finished, we took care of the children while Limor went out on her own board. I hadn't noticed that she was beginning to show until she expressed concern that she might be getting too big to surf any more until the new baby was born. Now there's a child out there somewhere who began amassing a critical kind of knowledge about a certain type of energy system before he or she was born. True knowledge. Bookmarks, models, secondary emotions, neural networks that accurately match the forces that play out over four fifths of the world. (It made me wish that I'd been able to fly co-pilot with my father during the war, before I was born. Or perhaps I did, and like Nick's son, I brought him back.)

Such early learning may be one of the most reliable kinds, forming deep knowledge and cool cognitions in high-energy states. It is learning that lasts a lifetime. All of the elite Superbike racers I came to know while working those races told me of experiences they'd had on motorcycles at a very early age. Miguel Duhamel was Honda's top racer. His mother told me that she rode while she was pregnant with him. When he was an infant and unable to sleep, she'd wrap Miguel in a bundle, get on the bike, and ride around the countryside. Miguel would quiet down. He loved riding as an infant. And he won more Superbike races than anyone else when I knew him. He was unbeatable, a tiny French Canadian jockey with a great screaming, 14,000-RPM horse.

As Mike Crowder and I watched Limor ride the faceted waves, we admired her natural grace and beauty. "Surfing is a connection with the universe," he said. "Surfing relaxes everybody." He smiled at the thought. Limor looked like Botticelli's *Birth of Venus* out there, drawing the *mana* into her womb from the sea, filling herself with that energy. Mike laughed as the sun angled across in shimmering waves. "Thank God for surfboards," he said.

TWO

SURVIVAL

Does anything happen to me?
I take what comes . . .

—Marcus Aurelius

On the occasion of every accident that
befalls you, remember to turn to yourself
and inquire what power you have for
turning it to use.

—Epictetus

NINE

BENDING
THE MAP

WHEN KEN KILLIP set out on the trail at Milner Pass in Rocky Mountain National Park at dawn on August 8, 1998, he had the nagging sense that he should not have come. A group of friends had planned the three-day backcountry hiking and fishing trip, but the others had gradually dropped out until only Killip and his friend John York were left. Killip, a firefighter, wondered if he should drop out, too, but decided to go ahead with the trip. From the trailhead, their route would follow the Continental Divide south for four miles, climbing 2,000 feet to the top of Mount Ida. There, at an altitude of 12,889 feet, they'd turn east, descend into the Gorge Lakes drainage, and hike two miles to Rock Lake. While six miles doesn't sound like much, hiking with a full pack to nearly 13,000 feet is serious business. In addition, Rock Lake sits at the edge of Forest Canyon, a densely wooded wilderness in the Big Thompson River valley. And as the local district ranger would later say, "It's one of the most remote areas in the park. It's pretty unforgiving."

Killip had plenty of outdoor experience. He had been with the Parker Fire Protection District just south of Denver for twenty-four years. He'd even had some survival training in the military. But Killip had never been in a place quite so rugged as this. And now the terrain, the altitude, and the heavy pack were taking their toll. He'd already given the tent to York to carry. York, a fellow firefighter and strong outdoorsman, repeatedly had to wait for Killip to catch up, and after five or six hours of that, York grew impatient and left Killip to fend for himself. Mismatching the abilities of people in the outdoors is a sure way to get into trouble. People routinely fail to realize that they have to travel at the speed of the slowest member, not the fastest.

Killip had been following York, who had been there before and knew the way. And although Killip had the map, York had the compass. They'd begun on a trail, but beyond the top of Mount Ida, it was a trailless wilderness, where you need both map and compass. Now, as he watched York disappear into the approaching weather, Killip didn't comprehend the insidious processes that were taking place. The world, though constantly changing, was the same as it had always been. The processes that would betray Killip were all taking place inside of him.

One type of mental model people form is a mental map: literally, a schematic of an area or a route. Killip had formed a sort of stochastic mental map of where he'd been since leaving his car. Because he'd been following York, he had not been checking his topographical map, and that is not a good way to create a reliable mental map. Now his brain was unconsciously trying to form a mental map of the route from a position he didn't really know to a destination he'd never seen before. That futile struggle contributed to his ill-defined anxiety.

In addition, a storm was rolling in, and Killip did not want to be the tallest object on the ridge. The area is well known for its afternoon lightning strikes. He decided to wait on the slope below the ridge until the thunderstorm passed. The multiple stresses of weather, fatigue, altitude, dehydration, and anxiety were closing in on Killip's ability to find that vital balance between useful emotion and reason.

As the storm began booming and flashing around him, it

increased his level of stress. Killip had time to reflect on his misgivings about the trip, but his thinking was already beyond
detailed, accurate analysis. In addition, stress was eroding his ability to perceive. He saw less, heard less, began to miss important
cues from his environment.

Four day-hikers came off the trail to join him and wait out the
rain. They told Killip that they'd seen his friend. Grasping at the
wished-for reality, Killip concluded that all he'd have to do was
hurry on ahead, and he'd find York. When at last the lightning
stopped, Killip pressed on in a driving rain, intent on salvaging the
trip.

He was climbing a steep slope that he was sure must be Mount
Ida—it just had to be. He'd been walking all day under his heavy
pack. He knew he would soon get to head down toward a cool,
clear river with a string of jewel-like lakes. He could drink. The
perception that he was climbing Mount Ida gave a more settled
feeling to the area of his brain that was trying to create a mental
map. At last, the hippocampus had something to work with. Killip
could picture Mount Ida and its relationship to his destination, and
mental maps are images. Without images, we are lost.

He'd been in motion for more than twelve hours. It was after
5:00 P.M., and he'd drunk the last of his water at about two o'clock.
The sun was going down; the temperature was dropping. The rain
continued to torment him.

When Killip struggled to the top, he turned east and began the
descent into the drainage, following his image of where Rock
Lake should be. But he immediately knew that something was
wrong: There was an unpleasant jolt from the amygdala. This was
not the place. The river and the little lakes and the rock shelf that
York had told him he would see weren't there. The image and the
world didn't match.

Killip had not, in fact, reached the summit of Mount Ida. He
was looking instead down a parallel drainage about a mile to the
north. Killip now teetered on the invisible dividing line between

two worlds: He was in a state of only minor geographical confusion. He could have retraced his steps. He still had a grip on one route, but he didn't have the big picture. He knew what was behind him. He did not know what was ahead of him. He could see into his past, but he had lost that vital cortical ability to perceive the world and therefore to see into his own future.

PSYCHOLOGISTS WHO study the behavior of people who get lost report that very few ever backtrack. (The eyes look forward into real or imagined worlds.) In Killip's case, there were other factors, too. He'd walked all day, exhausted, dehydrated, cold, and wet, probably by now feeling like a fool in York's eyes. He'd come a very long way, and his gut told him that it would be a long and painful way back, which would not lead to water. Rock Lake (and rest and water) had to be close at hand. If he'd been able to reason more clearly, he could have understood that he was not on the route to Rock Lake. But logic was rapidly being pushed into the background by emotion and stress. So, by the simple act of putting one foot in front of the other, he was about to cross over from mild geographical confusion to a state of being genuinely lost.

Edward Cornell, one of the scientists who study the behavior of people who become lost, is a professor of psychology at the University of Alberta in Edmonton. "Being lost is a universal human condition," he told me. "But there is a very fuzzy area between being lost and not lost."

Until about half a century ago, there was a widespread belief among scientists that people had some sort of inherent sense of direction. The observation that certain peoples around the world were especially skilled at navigation in the absence of obvious cues was evidence for a magnetic sense. The Australian Aborigines and the Puluwat Islanders in the South Pacific were examples of peoples who seemed inexplicably good at navigating. But when they studied those peoples more closely, researchers realized that they had simply been trained from childhood to pick up very subtle cues from the environment and use them the way anyone else would use landmarks to find a route. Even those people can and do get lost. And after half a century of research, it turns out that their greatest skill lay in keeping an up-to-date mental map of their environment.

There is no agreement among scientists on an exact definition of being lost. William G. Syrotuck, a pioneer in the field, defined it as being the subject of a land search. But many land searches are initiated for people who are just not where they're supposed to be. Kenneth Hill, a teacher and psychologist who also manages search and rescue operations in Nova Scotia, built on Syrotuck's work. He defines being lost as "30 minutes of not knowing where you are." That would suggest that a number of pilots I know have been lost in local taverns. Scientists who study human spatial cognition define being lost as being unable to relate your position in space to known locations. But being lost includes a whole range of emotional and behavioral consequences as well.

Syrotuck was the first search and rescue expert to conduct systematic research on the behavior of people who become lost in the wilderness. In *Lost Person Behavior*, he writes that they tend to panic. "Panic usually implies tearing around or thrashing through the brush, but in its earlier stages it is less frantic. . . . It all starts when they look about and find that a supposedly familiar location now appears totally strange, or when they start to realize that it seems to be taking longer to reach a particular place than they had expected. There is a tendency to hurry to 'find the right place.' 'Maybe it's just over that little ridge.'"

Recent research in neuroscience has shed some light on how people navigate. The way we know where we are is complex, as are the parts of the brain we use—the hippocampus and its component parts (such as the subiculum, the entorhinal cortex, and CA-3 and CA-1 formations). Joseph LeDoux calls the hippocampus "a spatial cognition machine." Neuroscientists have described how the brain creates mental maps of the environment. Early research with rats in the 1970s by John O'Keefe at McGill University, among others, provided the first neurophysiological evidence that the hippocampus creates "a spatial reference map" in the brain. In addition, there are cells that fire depending on the position of the head and others that track the position of the whole body or its parts. Still other cells fire only when traveling in one direction.

O'Keefe more or less accidentally found what he called "place cells" in the rat hippocampus. Place cells are individual neurons that get mapped to fire when the animal is at a specific place. Normally, hippocampal cells fire perhaps only once every second on average. But at that mapped place, they fire hundreds of times faster. In tests with monkeys at the University of Oxford, cells were found that fired only when the animal was looking at a certain view. A single cell can map more than one place.

So there is an elaborate system involving the hippocampus and other areas of the brain for creating an analog of the world and your motion, position, and direction of travel within it. It works in concert with other systems to locate you in your mind. For example, through information from the inner ear (the vestibular system), your brain is constantly telling you whether you're upright or not and whether you're leaning over or falling backward. Through the proprioceptive system, the brain is constantly reading signals from nerve cells throughout the body to tell you where the parts of your body are. That's why you can touch your nose with your eyes closed. Without those systems, you'd get lost every time you tried to go anywhere, as experimental animals do when researchers destroy the hippocampal area of the brain. Alzheimer's patients do, too.

Place cells and other cells involved in navigation are constantly being reprogrammed. It's called "remapping." Any time you go to a new place, the brain begins trying to create a new map. For some people, it takes only one trip; others have to repeat the route several times to remember it. (We've all had the experience of waking in a strange place and not knowing where we are.) The hippocampus is associated with memory, and the maps appear to be stored in the same way as memory. You create not just routes but maps of areas of your environment, such as a room, your house, or your whole neighborhood. Many people find, for example, that they can easily navigate around their own bedroom or even large parts of their house without the lights on, because the mental map in their brain matches the real world. Blind people often get around just fine because they have excellent mental maps. Place cells in rats fire in the dark. But stress interferes with the work of the hippocampus, making it harder to make and revise your mental maps.

Interestingly, the hippocampus, which tells you where you are and where you're going (if the map is right), does not control the seeking of a goal. The urge to get to a specific place, the drive toward a goal, appears to be emotional. That makes sense, since the amygdala helps trigger action, especially as it relates to survival. Rats who have had the lateral nucleus of the amygdala destroyed lose their drive to get to a particular place. So, place and motivation are integrally connected, which may explain what keeps people moving when it would be safer for them to stay still.

When a person goes from a city to, say, Rocky Mountain National Park, it puts some unusual demands on the brain. In the city, all the visual cues are near and limited in number as well. You may see the inside of your home, the inside of your office, streets bordered by buildings, and so on. Rarely do most people get a sweeping panorama in a city. When you travel to the mountains, suddenly all the cues are different, as are all the requirements of mapping that are going on continuously and unconsciously in the brain. The brain is reaching out through the senses, bringing information in, attempting to grasp the environment and wire up a map. The input and output of the hippocampus and other areas are being sent to the amygdala to establish a drive toward beneficial things and an aversion to harmful things. The amygdala is set to respond with action. For a person displaced from his normal environment, the task of mapping the unfamiliar and vast world might feel a bit overwhelming.

SO, WITH REASON pretty much out of the picture and emotion driving hard toward survival strategies, Killip started down the wrong drainage as darkness and rain fell around him. The absence of a mental map of the place in which he found himself caused the amygdala to begin sending signals of danger. People recognize as good such places as the location of food, water, and members of the opposite sex. That's a primary task of adaptation and survival. People also recognize dangerous places. And it makes perfect sense that a dangerous place to be is one for which you have no mental map, for then you'd be unable to find food, water, or a mate.

Killip's seemingly irrational behavior makes sense when viewed from the brain's point of view. The fact of not having a mental map, of trying to create one in an environment where the sensory input made no sense, is interpreted as an emergency and triggers a physical (i.e., emotional) response. In the emergency of being no place, Killip's action makes sense to the organism, even though it later seems illogical. The organism needed him to hurry up and try to get some place quickly, a place that matched his mental map, a place that would provide access to the essentials of survival. This impulse explains Syrotuck's observation that people panic when they become lost. It gives a working definition of being lost: the inability to make the mental map match the environment.

So it was that Killip found himself blundering through dense timber in total darkness with the creepy feeling of knowing that he was nowhere. A chance flicker of lightning ignited reflections on a pond. Parched with thirst, Killip headed for it. He drank his fill and prepared to spend the night. He had no choice now. But he wasn't thinking straight. He had food in his pack, but York had the tent. Killip had garbage bags but didn't use them for a makeshift shelter. Although he needed a fire, wanted its warmth and light, he knew that open fires weren't permitted in this part of the park. As a firefighter, he felt he ought to follow that rule. (If he had made a fire, he might have been seen and rescued sooner.)

When a bear appeared, Killip got up and charged the animal, waving his jacket at it and shouting. The bear went away. Then Killip wondered what would have happened if he'd been injured so far from help.

He was able to heat a meal on his camp stove. Then he fell asleep.

When he awoke, he felt somewhat refreshed. But he would not recover from his fatigue and confusion that quickly. He still had the option of retracing his steps to his car. He could go back up the drainage. But he felt that he could not simply leave York and spoil the trip. York would be thinking: What a nitwit. And anyway, Killip didn't yet quite believe that he was lost.

Admitting that you are lost is difficult because having no mental map, being no place, is like having no self: It's impossible to conceive, because one of the main jobs of the organism is to adjust itself to place. That's why small children, when asked if they are lost, will say, "No, my Mommy is lost." The sense is: I'm not lost; I'm right here. But without a mental map, the organism can't go about its business and rapidly deteriorates. So to Killip, it seemed that he wasn't lost. Rock Lake was lost. It had to be just around the corner somewhere. Then everything would be all right. He had a firefighter's can-do persistence and a lost person's tendency to form a strategy, albeit a faulty one.

He began bushwacking through forest so dense he sometimes had to remove his backpack to squeeze between the trees. It didn't occur to him that this might be a bad sign. But anytime you find yourself thinking it's easier to go around a mountain than over one, you know there's trouble upstairs.

As Syrotuck writes, "If things get progressively more unfamiliar and mixed up, [the victim] may then develop a feeling of vertigo, the trees and slopes seem to be closing in and a feeling of claustrophobia compels them to try to 'break out.' This is the point at which running or frantic scrambling may occur," as the organism frantically attempts to get a fix on an alien environment.

By afternoon, Killip's wanderings had severed all connection to the world he'd known. His circle of confusion had expanded so that he could no longer even retrace his steps: He was profoundly lost. And while the rational part of his brain remained convinced that he was getting close to Rock Lake, the emotional part was driving him on with more and more urgency. (He'd eventually pass within a quarter mile of Rock Lake, but not that day.)

As his brain continued the unconscious search for any cue with which to establish a mental map, the alarm signals grew more and more urgent: No place. No food. No people. *Get there, get there,* a voice seemed to say. *Hurry, hurry.*

Killip began scrambling up a steep scree slope to get a better view. Myabe if he just got up high . . . if he could just see the whole area, then everything would snap back into focus and he could calm down.

About halfway up, he lost his footing and couldn't self-arrest. He began to cartwheel down the long grade. When he came to a stop, he had suffered severely pulled muscles in his shoulder, ligament and cartilage damage in his knees, and two sprained ankles. He was lucky it wasn't fatal, as such a violent fall often is.

Killip dragged himself to a small pond, where he had no choice but to remain through another rainy night. He tried to reason, but something was wrong. He was cold and hurt but still believed he was forbidden from making a fire. He didn't even erect a shelter.

Killip awoke in pain and frustration. He'd had it with trying to find Rock Lake. He was definitely going back to the car, he decided. Although he had no idea what direction to go, because he didn't know where he was, he began limping through the forest, battering his way through the trees, wasting precious energy. But with no understanding of what was happening to him, he could not settle down. Once again, he decided that the best strategy was to climb up and see if he could get an overall view of where he was.

As Syrotuck put it, "If they do not totally exhaust or injure themselves during outright panic, they may eventually get a grip on themselves and decide on some plan of action. What they decide to do may appear irrational to a calm observer, but does not seem nearly so unreasonable to the lost person who is now totally disoriented. Generally, they would be wiser and safer to stay put and get as comfortable and warm as possible, but many feel compelled to push on, urged by subconscious feelings." Urged on by the frustrating task that his unconscious brain activity had been trying to complete for so long now without success. The organism's main task is to map the self, map the environment, and keep the two in harmonious balance. Without the balance, the organism dies.

Killip began struggling up another steep and rocky slope. It was actually Terra Tomah Mountain, a 12,718-foot peak. But before he could get himself rim-rocked, a storm blew in and forced him back down toward the trees. He felt woozy. He felt strange. He knew he was in serious trouble, but there didn't seem to be anything he could do about it. He passed out with one arm slung around a tree trunk to keep himself from sliding down the steep rock.

It was past midnight when he awoke, wet and shaking uncontrollably. He looked around. The world was strange. Everything was white. After a moment, he realized what he was seeing: Hailstones covered the ground to a depth of 12 inches. He had slept through a big storm.

When he'd set out on August 8, Killip had been a healthy, competent, well-equipped hiker. His pack contained everything he needed to survive at least a week in the wild. Now, just over two days after taking a wrong turn off the Continental Divide, he was huddled on an icy mountainside, exhausted, hungry, badly dehydrated, injured, and dangerously hypothermic. What had begun as a small error in navigation had progressed, step by innocent step, to a grim struggle for survival.

. . .

SYROTUCK ANALYZED 229 search and rescue cases (11 percent of them fatal) and concluded that almost three quarters of those who died perished within the first forty-eight hours of becoming lost. Those who die can do so surprisingly quickly, and hypothermia is usually the official cause. Hypothermia is frighteningly insidious, but in some cases people just give up.

Anyone can get lost. I know. I have. But surprisingly few are genuinely prepared to live through the experience. I was staying at Many Glacier Hotel, in Glacier National Park, and decided to hike the half-hour nature trail with a friend before breakfast. But there was the air, and the view, and that spicy juniper smell of the mountains. There were those dizzying spaces and the Hansel and Gretel forest beckoning . . . Our experience of a week of hard hiking deep in the Montana wilderness had convinced us both that we knew our way around. We were fit and confident.

We left the little loop trail and followed a sign for Grinnell Lake. We took one fork, then another, then another. When the first drops of rain started falling, we slipped into our cheap gift-shop ponchos and hurried on with a growing sense of urgency. We didn't consider turning back.

At last, we stood on the shore of a lake, trying to remember why it had seemed so important to get there. The soft hissing of rain suddenly accelerated to a clattering of hailstones. I looked over at my companion and saw that her face was pale and blotchy. Her teeth had begun to chatter. I felt a cold dread set like plaster in my stomach as the realization hit me that we were standing in a hailstorm, dressed in cotton T-shirts and garbage bags, at least two hours from home with no map or compass. *What were we thinking?*

We took off down the trail at a dead run, but when we reached a fork, she went one way and I went the other. We turned back to look at each other in horrified amazement. We had no idea which way we'd come. We finally chose one of the paths and had just set foot on it when we heard a human voice. We surged toward it, crashing through a few yards of dense forest, and found ourselves at a dock on another lake, where a tourist boat had just pulled up. Ice had already covered the windshield.

As we clambered aboard, we were told that it was the last boat of the day. I still sometimes wake up at night wondering what it would have been like if we had stayed on that path (which led deeper into the wilderness, as we later discovered from a glance at the map we'd left at the hotel) with no water, no fire, and no warm clothing, in what turned out to be a two-day ice storm.

In just a few hours, we'd gone from being carefree day hikers to panicked victims, saved only by dumb luck. Until that day in Glacier, I would not have believed how easily I could get lost or how quickly I could lose my ability to reason.

One of Kenneth Hill's experiments involves taking a group of his students into a small forest in Nova Scotia. "It's about the size of a large city park," he told me, "and notorious for its maze of poorly marked trails." He leads the students in and then asks them to lead him out. Only one person has ever succeeded. "If you ask hikers on a trail to point out where they are on a map at any given moment," Hill said, "they are usually wrong."

In daily life, people operate on the necessary illusion that they know where they are. Most of the time, they don't. The only time most people are not lost to some degree is when they are at home. It's quite possible to know the route from one place to another without knowing precisely where you are. That's why streets have signs. Nevertheless, most people normally have enough knowledge of a route to get them where they're going. If they don't—as in my case, and in Ken Killip's case—they get lost.

It's simple. All you have to do is fail to update your mental map and then persist in following it even when the landscape (or your compass) tries to tell you it's wrong. Edward Cornell once told me, "Whenever you start looking at your map and saying something like, 'Well, that lake could have dried up,' or, 'That boulder could have moved,' a red light should go off. You're trying to make reality conform to your expectations rather than seeing what's there. In the sport of orienteering, they call that 'bending the map.'"

Killip was bending the map when he headed down the wrong drainage despite ample evidence that he was starting from the wrong place. But it's understandable how urgent it feels to make your mental map and the world conform. It's the essence of what every organism does, even those that don't have cognition as we know it.

If we persist in bending the map until we can no longer deny the evidence of our senses, it can be terrifying. "It's not something that happens immediately," Hill said. "First, it's a sense of disorientation: 'Uh-oh, I'm not in Kansas anymore.' Then the woods start to become strange; landmarks are no longer familiar."

Since the organism's survival depends on a reasonable match between mental map and environment, as the two diverge, the hippocampus spins its wheels and the amygdala sends out alarm signals even as the motivational circuits urge you on and on. The result is vertigo, claustrophobia, panic, and wasted motion. Since most people aren't conscious of the process, there's no way to reflect on what's happening. All you know is that it feels as if you're going mad. (And what else is insanity but a failure to match mind and world?) When at last the full weight of the incongruity hits you, the impact can be devastating. (Psychologists have observed that one of the most basic human needs, beginning at birth, is to be gazed upon by another. Mothers throughout the world have been observed spending long periods staring into the eyes of their babies with a characteristic tilt of the head. To be seen is to be real, and without another to gaze upon us, we are nothing. Part of the terror of being lost stems from the idea of never being seen again.)

People have known for ages that going from the protection of society into the wild can have a profound effect on the balance of reason and emotion. It can induce altered states of consciousness, hallucinations, even death. The word "bewildered," with its definite, familiar Anglo-Saxon ring, dates from 1684 and comes from the archaic verb, "wilder." To "wilder" someone means to lead him into the woods and get him lost. But a recent Webster's dictionary definition retains much of the original Old English sense:

Bewilder, *v.t.*; bewildered, *pt.*, *pp.*; bewildering, *ppr.* [Dan. *forvilde*, to bewilder; G. *verwildern*; AS. *wilde*, wild.]
1. to confuse hopelessly; befuddle; puzzle.
2. to cause (a person) to be lost in a wilderness. [Archaic.]
Syn.——daze, dazzle, confound, mystify, puzzle, astonish, perplex, confuse, mislead.

Bewilderment, *n.* 1. the fact or state of being bewildered; a chaotic state of the mental forces; perplexity.

Wild, *a.* [ME. *wilde*, *wielde*, from AS. *wild*, wild, bewildered, confused.]

The more modern term "woods shock," which is used by psychologists, dates from at least 1873, where it appeared in the journal *Nature*. It refers to a state of confusion that can beset people in the wilderness.

" 'Woods shock' is a term for the fear associated with complete loss of spatial orientation," Kenneth Hill told me. "None of the rational abilities that the victim had before being lost are useful to him anymore." In severe cases, the actions of even the most experienced outdoorsmen can seem inexplicable. Hikers have abandoned full backpacks; hunters have left their guns behind. Killip neglected to make fire or shelter.

But in the light of recent advances in neuroscience, woods shock can now be seen as an emotional survival response associated with the failure of the mental map to match the environment. Thrashing does not save a drowning person either, but it's just as natural. Those who can float quietly have a better chance.

EVERYONE WHO dies out there dies of confusion. There is always a destructive synergy among numerous factors, including exhaustion, dehydration, hypothermia, anxiety, hunger, injury. So woods shock, which can now be explained in the more precise terms of neuroscience, led Ken Killip to frantic, poorly planned actions. Those stresses and actions incapacitated him even further in a tightening spiral until reason and emotion, instead of working in harmony to produce correct action, became like two drowning swimmers, dragging each other down.

Being lost, then, is not a location; it is a transformation. It is a failure of the mind. It can happen in the woods or it can happen in life. People know that instinctively. A man leaves a perfectly good family for a woman half his age and makes a mess of it, and people say, he got off the path; he lost his way. If he doesn't get back on, he'll lose the self, too. A corporation can do the same thing.

The research suggests five general stages in the process a person goes through when lost. In the first, you deny that you're disoriented and press on with growing urgency, attempting to make your mental map fit what you see. In the next stage, as you realize that you're genuinely lost, the urgency blossoms into a full-scale survival emergency. Clear thought becomes impossible and action becomes frantic, unproductive, even dangerous. In the third stage (usually following injury or exhaustion), you expend the chemicals of emotion and form a strategy for finding some place that matches the mental map. (It is a misguided strategy, for there is no such place now: You are lost.) In the fourth stage, you deteriorate both rationally and emotionally, as the strategy fails to resolve the conflict. In the final stage, as you run out of options and energy, you must become resigned to your plight. Like it or not, you must make a new mental map of where you are. You must become

Robinson Crusoe or you will die. To survive, you must find your-self. Then it won't matter where you are.

The stages of getting lost apply to more than just hiking in the woods. A company, such as Xerox, ignores cues from a changing world and from inside its own research facility in Palo Alto and nearly destroys itself. In 1959, Xerox introduced its 914 copier. *Fortune* said it was "the most successful product ever marketed in America." By 1969, Xerox passed $1 billion in sales. In 1971, flushed with success (an emotional state of high arousal), the com-pany's officers were in a state of deep denial. The world was changing, and they weren't taking in any new information. Their cup was full. At the stockholders' meeting that year, these words, which would nearly destroy the company, were uttered: "We can handle all your information needs." Xerox's leaders had decided to take on IBM, despite all the clear evidence that it would most likely kill them to do so. They were like the snowmobilers, flushed with emotion, who went up that hill, despite the clear evidence that it would probably kill them. They were bending the map, too.

Xerox spent $1 billion to purchase a computer company. At the same time, the company opened the Palo Alto Research Center (PARC). It took only five years (which is like five days for a hiker) for the computer business to drag Xerox down by about $85 mil-lion in cash. In the meantime, scientists and engineers at PARC were inventing the mouse, Ethernet, the graphical user interface, the flat-panel display, and the laser printer. Others got rich off of those inventions. Xerox, busy with its mental models that did not match the real world, saw none of that profit.

Unlike Ken Killip, Xerox is still in the woods.

The stages of getting lost resemble the five stages of dying described by Elisabeth Kübler-Ross, the psychologist who wrote *On Death and Dying*: denial, anger, bargaining, depression, and acceptance. The end result is often the same. "Once the stage of psychological disintegration is reached, death is often not far away," John Leach writes in *Survival Psychology*. "[T]he ability people possess to die gently, and often suddenly, through no organic cause, is a very real one." This suggests that some cases listed as hypothermia may not have been.

That's a lesson Kenneth Hill knows well. "I have photos of a man who settled into a cozy bed of pine needles after removing his shoes, pants, and jacket and setting his wallet on a nearby rock," he told me. "In the photos, he seems so peaceful; it's hard to believe he's dead. The photos have special significance for me, because I helped coordinate the search. Whenever I start to believe I'm some hot-shit SAR expert, I pull the photos out and I'm over it."

Consciousness is a murky, intermittent phenomenon that has yet to be debugged. It sees the world through a glass darkly, not face to face, as Paul the Apostle said. Many conditions influence how you perceive and how much you perceive, as well as what you do with that information. So you go unconsciously about your business, losing your keys and finding them right under your nose. Running red lights. Letting the pot boil over. Forgetting to pay the electric bill. The consequences are few. Then you go into the wilderness, where the consequences are many.

BY HIS third night lost in the wild, when Killip awoke amid the hailstones at the foot of Terra Tomah Mountain, he had arguably passed through the stages of denial (descending the wrong drainage), panic (climbing up the dangerous scree slope), and strategic planning (attempting to backtrack), and was well into the penultimate stage of deterioration. But he did not succumb to resignation.

That happens in a lot of cases (including big companies, troubled marriages, sick people, lost souls). There are great survivors and helpless victims on the curve of human ability. Most of us are neither. Most of us fall somewhere in between and may perform poorly at first, then find the inner resources to return to correct action and clear thought. If the object of the game is survival, that

will do. Or, as my father used to tell me about flying, "A good landing is any landing that you can walk away from."

Killip pulled himself together. He put on his fishing waders and
started walking around to get warm. He made a fire and built a
makeshift shelter using his garbage bags. (Both were things he
should have done the first day, but better late than never.) For the
next two days, he stayed put and attended to the business of adapting to the environment, keeping the organism in balance, the
process called survival. Killip had entered the final stage that separates the quick from the dead: not helpless resignation but a prag-
matic acceptance of—and even wonder at—the world in which he
found himself. He had at last begun to model and map his real
environment instead of the one he wished for. He'd worked out his
own salvation. He had discovered the first Rule of Life: *Be here now*.

That final stage in the process of being lost can prove to be
either a beginning or an end. Some give up and die. Others stop
denying and begin surviving. You don't have to be an elite performer. You don't have to be perfect. You just have to get on with it
and do the next right thing.

Having faced the reality of his situation, having created a mental map of where he was, not where he wanted to be, Killip was
now able to function. On his fifth and final day, he watched as a
helicopter passed right over him, so close that "I felt like I could
throw a rock at him. Then it turned and flew away. It was almost
breaking my spirit."

One of the toughest steps a survivor has to take is to discard the
hope of rescue, just as he discards the old world he left behind and
accepts the new one. There is no other way for his brain to settle
down. Although that idea seems paradoxical, it is essential. I know
that's what my father did in the Nazi prison camp: He made it his
world. Dougal Robertson, who was cast away at sea for thirty-eight
days, advised thinking of it this way: "Rescue will come as a welcome interruption of . . . the survival voyage."

The helicopter pilot had seen Killip's blue parka hanging on a branch and directed the ground searchers to his location. "I lost thirty pounds in five days," Killip told me. His knee injuries required two operations. Today, he still goes into the wilderness, but "now I carry a survival pack and a map and compass every- where. And I'm very careful about who I go out with. If I have a bad feeling about something, I don't go."

ONE OF the many baffling mysteries concerns who survives and who doesn't. "It's not who you'd predict, either," Hill, who has studied the survival rates of different demographic groups, told me. "Sometimes the one who survives is an inexperienced female hiker, while the experienced hunter gives up and dies in one night, even when it's not that cold. The category that has one of the highest survival rates is children six and under, the very people we're most concerned about." Despite the fact that small children lose body heat faster than adults, they often survive in the same conditions better than experienced hunters, better than physically fit hikers, better than former members of the military or skilled sailors. And yet one of the groups with the poorest survival rates is children ages seven to twelve. Clearly, those youngest children have a deep secret that trumps knowledge and experience.

Scientists do not know exactly what that secret is, but the answer may lie in basic childhood traits. At that age, the brain has not yet developed certain abilities. For example, small children do not create the same sort of mental maps adults do. They don't understand traveling to a particular place, so they don't run to get somewhere beyond their field of vision. They also follow their instincts. If it gets cold, they crawl into a hollow tree to get warm. If they're tired, they rest, so they don't get fatigued. If they're thirsty, they drink. They try to make themselves comfortable, and staying comfortable helps keep them alive. (Small children follow- ing their instincts can also be hard to find; in more than one case, the lost child actually hid from rescuers. One was afraid of "coyotes" when he heard the search dogs barking. Another was afraid of one-eyed monsters when he saw big men wearing headlamps. Fortunately, both were ultimately found.) The secret may also lie in the fact that they do not yet have the sophisticated mental mapping ability that adults have, and so do not try to bend the map. They remap the world they're in.

Children between the ages of seven and twelve, on the other hand, have some adult characteristics, such as mental mapping, but they don't have adult judgment. They don't ordinarily have the strong ability to control emotional responses and to reason through their situation. They panic and run. They look for shortcuts. If a trail peters out, they keep going, ignoring thirst, hunger, and cold, until they fall over. In learning to think more like adults, it seems, they have suppressed the very instincts that might have helped them. But they haven't learned to stay cool. Many may not yet be self-reliant. (One of my survival instructors told me that inner-city children did better in survival training than ones from the suburbs, "because the suburban children have no predators.") They have begun to learn to navigate, to make detailed mental maps; children that age trade secret routes and shortcuts. But a little knowledge is dangerous. A child that age will run across a road without stopping when lost.

We like to think that education and experience make us more competent, more capable. But it seems that the opposite is sometimes true. Ultimately, after years of chasing the ghost of the boy aviator who went on to become my father, I went to survival school to try to find out if there was a way to learn the skills of survival while retaining the instincts of a small child. I thought that perhaps through all the risk taking, the quest for cool, I had wandered off the path of life. I couldn't help thinking, then, of the Zen concept of the beginner's mind, the mind that remains open and ready despite years of training. "In the beginner's mind there are many possibilities," said Zen master Shunryu Suzuki. "In the expert's mind there are few."

TEN

INSIDE THE
RIGHT STUFF

JULIANE KOEPCKE WAS flying with her mother and ninety other passengers on Christmas Eve, 1971, when lightning struck, causing an extensive structural failure of the Lockheed Electra. Later, she reported feeling "a hefty concussion." Then she was falling toward the jungle. Juliane fell out of the broken airplane into the Peruvian jungle. She was seventeen years old, wearing her Catholic confirmation dress and white high heels. Miraculously, she suffered only cuts and a broken collarbone from the crash.

As she recalled, "I remember thinking that the jungle trees below looked just like cauliflowers." To someone who knows about survival, that statement is telling. She wasn't screaming; she wasn't in a panic. She was in wonder at the world in which she found herself. She was taking it all in, touching her new reality. Checking out her environment while falling. Amazing cool.

Amazing and also characteristic of a true survivor. Bill Garleb, an American GI who survived the Bataan Death March in the Philippines, found his senses increasingly sharp as he experienced a deep wonder at the birds and colors and smells of the jungle.

A dozen other passengers survived the midair disintegration of

Juliane's plane, and their attitude, and hence their behavior and fate, were quite different from hers.

Juliane awoke alone on the floor of the jungle, still strapped into her seat. There was no sign of her mother, who'd been beside her in the plane. She spent the night trying to keep out of the rain under her seat. The next day, she deduced that even the helicopters and airplanes she could hear wouldn't be able to see her through the jungle canopy. She'd have to get herself out. It was another important moment: She didn't spend time bemoaning her fate. She looked to herself, took responsibility, made a plan.

Her parents were researchers who worked in the jungle, and she was familiar with that environment. But Juliane had had no survival training. She didn't know where she was or which way she ought to go, but her father had told her that if she went downhill, she'd find water. He'd said that rivers usually led to civilization. And while that strategy can just as easily lead into a swamp, at least she had a plan that she believed in. She had a task.

Meanwhile, the others who had lived through the fall decided to await rescue, which is not necessarily a bad idea either. But expecting someone else to take responsibility for your well-being can be fatal. In *Alive*, Piers Paul Read tells the story of the survivors of another airplane crash, this one in the Andes. Everyone who survived the crash stayed put, assuming that they'd be rescued. Many died; the others wound up eating each other to keep from starving before someone finally walked out and found help.

Juliane had nothing except a few pieces of candy and some small cakes. She had no survival equipment, no tools, no compass or map—none of the things I'd been taught to use in survival school. But she very deliberately set up a program for herself. She set off, resting through the heat of the day and traveling during the cooler periods. She walked for eleven days through dense jungle while being literally eaten alive by leeches and strange tropical insects, which bored into her, laid their eggs, and produced worms that hatched and tunneled out through her skin.

Eventually, she came to a hut along the banks of the river she'd been following. She staggered and collapsed inside. There is always a lot of chance involved in a survival situation, both good luck and bad. It was Juliane's good fortune that three hunters turned up the next day and delivered her to a local doctor. But, as Louis Pasteur said, "Luck favors the prepared mind."

Tough and clearheaded, this teenage girl, who had lost her shoes (not to mention her mother) on the first day, saved herself; the other survivors took the same eleven days to sit down and die.

The forces that put them there were beyond their control. But the course of events for those who found themselves alive on the ground were the result of deep and personal individual reactions to a new environment.

The knottiest mystery of survival is how one unequipped, ill-prepared seventeen-year-old girl gets out alive and a dozen adults in similar circumstances, better equipped, do not. But the deeper I've gone into the study of survival, the more sense such outcomes make. Making fire, building shelter, finding food, signaling, navigation—none of that mattered to Juliane's survival. Although we cannot know what the others who survived the fall were thinking and deciding, it's possible that they knew they were supposed to stay put and await rescue. They were rule followers, and it killed them.

In the World Trade Center disaster, many people who were used to following the rules died because they did what they were told by authority figures. An employee of the Aon Insurance Company on the ninety-third floor of the south tower had begun his escape but returned to his office after the security guards made a general announcement that the building was safe and that people should stay inside until they were told to leave. Before he died, he spoke to his father on the phone: "Why did I listen to them—I shouldn't have." Another man, an employee of Fuji Bank, actually reached the ground-floor lobby, only to be sent back in by a security guard. A third worker called a family member and recorded a final message on the answering machine: "I can't go anywhere because they told us not to move. I have to wait for the firefighters."

In thinking for herself, Juliane wasn't even particularly brave. Survival is not about bravery and heroics. Heroes can be perfect heroes and wind up dead. By definition, survivors must live. Juliane was afraid most of the time (of everything from piranhas when she had to wade in the water to the worms that were crawling around under her skin to the real or imagined creatures of the forest). Survivors aren't fearless. They *use* fear: they turn it into anger and focus.

Conversely, searchers are always amazed to find people who have died while in possession of everything they needed to survive. John Leach writes that "Victims have been recovered from life rafts with a survival box (containing flares, rations, first-aid kit and so on) unopened and the necessary contents unused."

"Some people just give up," Ken Hill told me, referring to his search and rescue operations in Nova Scotia. "Fifteen years I've been studying this, and I can't figure out why."

What saved Juliane was an inner resource, a state of mind. She certainly didn't have any physical equipment. But she'd been prepared mentally, somehow. A lifetime of experience shapes us to meet or be crushed by such challenges as a bad divorce, the shattering of a career, a terrible illness or accident, a collapsing economy, a war, prison camp, the death of a loved one, or being stranded in the jungle. I went to survival school to try to understand that mystery and see if I could master my own journey.

MY FRIEND Jonas Dovydenas was born in Kaunas, Lithuania, in 1939, and fled with his parents in 1944 to escape the Soviets. As Displaced Persons in Ingolstadt, Germany, they were bombed by my father's outfit, the Eighth Air Force. They left Germany in 1949 to come to the United States. When Jonas visits my father now, he wears a baseball cap that features the Air Force shield and looks just like my father's cap from his 398th Bomb Group reunions. Jonas's cap, however, says "Bombee" on the front.

Jonas studied writing and English literature at Brown, then photography at the Rhode Island School of Design and at the Design Institute at Illinois Institute of Technology in Chicago. In 1969, he introduced me to a magazine editor and got me my first assignment. He and I have been working and traveling together ever since—he takes the pictures, I write the words.

When I got my pilot's license, Jonas was my first passenger, and the bug bit him so hard that he built his own replica of a two-seater Italian fighter plane, the Falco, in which we've flown all over the United States. I like to travel with Jonas because he's a survivor. Like me, he's been after it since childhood. Jonas is cool. I named my son after him.

So, when I went to survivor school, it seemed natural to me that Jonas should come along. We took off from Pittsfield, Massachusetts, sharing the flying duties, and landed in Lynchburg, Virginia. As I climbed out of the cockpit and onto the wing, I caught my first glimpse of our instructor. Byron Kerns runs the Mountain Shepherd Survival School. He swaggered across the fueling ramp toward us wearing an 18-inch Panamanian machete on his belt, a big, macho-looking guy with twelve years of military experience, including a stint with the Marines. He had worked at the famous Air Force Survival School in Washington State. When I saw him, I thought: We're in for it now.

That night, Byron explained that we were going to head off into the Virginia woods the next morning, early, and we'd be drilling for several days on such matters as map and compass work, firecraft, shelter, and signaling. We'd learn to find water or distill it from the air. We would not think about food, because it wasn't necessary. The Air Force plan was that you'd be found within three days. "Your job," he said, "is to stay alive for seventy-two hours." When he left, Jonas said, "This guy's going to whip you like a red-headed stepchild."

Early the next morning, as we moved up a rocky river drainage through the mountains, I noticed that Kerns would stop frequently to point out something of beauty or interest. He spoke softly, as if we were in a church. He laughed a lot. He liked to be still and just think or smoke a cigarette. I saw no sign of the drill sergeant I'd expected. In our first exercise, Kerns asked me and Jonas to make a fire, and in a matter of minutes we had a roaring blaze going. Kerns had turned away to get something from his pack. When he turned back, the flames were leaping several feet off the ground. "Whoa," he said, laughing, "easy, easy. I just wanted to know if you could start a fire. Some people can't." Then he gently separated the pile of wood and put it out.

Byron Kerns turned out to be soft-spoken, polite, cheerfully earnest, and gentle to a fault. He moved slowly, never hurried, and was always carefully assessing himself and his environment. He wasn't prone to high emotional states. He carried with him a contagious air of calm. He reminded me of my father, actually. Like so many retired pilots, my father wore soft shoes, talked softly, and walked slowly. (As a pilot, you want to wear soft shoes so that you can feel the rudder pedals. You don't want to make sudden, unplanned motions in a combat aircraft cockpit, where the controls are sensitive and lots of things can explode.) That demeanor, once learned under penalty of death, is carried through the rest of your life. Kerns also had that quiet, dark, and private humor.

Even after a lifetime in the wilderness, Kerns entered the woods with a deep sense of respect and humility, like a man approaching a magnificent, dangerous, and unpredictable creature. It's the same way a good pilot approaches his aircraft.

As we worked in the wilderness, learning technical skills, Kerns kept talking about Positive Mental Attitude. It was the number one item on the checklist he'd given us, and that checklist was from the Big Daddy of all checklist writers, the U.S. Air Force. Positive Mental Attitude.

"It must be important," I told Jonas.

"Yeah, but what is it?" he asked.

"Think good thoughts and you'll be saved?"

"I'd rather have a chain saw and a cheeseburger," Jonas said. "A cell phone and a GPS would be nice, too."

As we slogged through the woods, practicing firecraft, shelter-making, knots, and navigation while the mammoth Virginia ticks gnawed on our privates, I kept asking Kerns, but he couldn't explain it. Nobody could. It meant the difference between life and death; he could tell me that. He had an adult portion of it; he assured me of that. "It's not what's in your pack," he'd say. "It's what's in here." He'd tap his chest. No wonder Tom Wolfe had called it *The Right Stuff*. You couldn't exactly title a book *Positive Mental Attitude*, now, could you?

Kerns didn't always have it, though. He'd had to acquire it. Early on in his Air Force days, he took a group of pilots into the mountains near Spokane for survival training maneuvers. "I was a greenhorn and just misjudged our situation," he told us. Back then, he was pretending to be the macho drill instructor I'd expected: Go, go, go, push, push, push. He was not yet cool. He was acting cool.

His class had been crossing a vast field of slushy snow, which made the going rough. The pilots began to suffer from fatigue, but Kerns kept driving them. "I now realize that was a mistake," he said. As the temperature dropped, darkness came down like a curtain. "Suddenly everybody wanted to give up. They just sat down and lost all their will." Apathy is a typical reaction to any sort of disaster, and if you're exhausted in a field of snow at sundown in the mountains, you're pretty much about to witness the simple disaster of nature separating you permanently from everything you know and love in this world. That apathy can rapidly lead to complete psychological deterioration. Then you sit down and hypothermia sets in, which produces more apathy, a more profound psychological deterioration, and ultimately, death.

Fatigue almost always comes as a surprise. It is as much a psychological condition as a physical one, and scientists have struggled without success to understand it. It's like the difficulty of studying sand in order to understand the Sand Pile Effect. There's nothing in the muscles or nerves or even the biochemistry of the body that would seem to predict or explain fatigue. Once fatigue sets in, though, it is almost impossible to recover from it under survival conditions. It is not just a matter of being tired. It's more like a spiritual collapse, and recovery requires more than food and rest.

Following the explosive burst of activity that is sometimes required for survival, or in the panic stage when you're running or climbing or swimming, you're like a woman who's just given birth to a baby. You're depleted and wide open to fatigue. It may take weeks to recover; and if you're not taking care of yourself, that fatigue can lead to an inability to sleep, which in turn can result in a sudden psychological collapse. The physical and psychological factors rapidly erode each other, which is why it is so important to pace yourself, rest frequently, and stay hydrated. That's why Kerns's pushing the pilots so hard had been a mistake.

A survival situation is a ticking clock: You have only so much stored energy (and water), and every time you exert yourself, you're using it up. The trick is to become extremely stingy with your scarce resources, balancing risk and reward, investing only in efforts that offer the biggest return.

In survival situations, people greatly underestimate the need for rest. While Kerns, Jonas, and I were doing map and compass exercises, he would frequently stop and look around at the woods, chatting with us. I'd be thinking: *Let's go, let's go, I know the way.* And he'd just stand there. Now I understand why. You should operate at about 60 percent of your normal level of activity, he explained, and rest and rehydrate frequently. If the weather is cool and you're sweating, you're working too hard.

I'd had the same lesson when I was training with Air Force fighter pilots to fly combat aircraft. I'd show up for the morning briefing, and when it was over, I'd be pacing around in my "speed jeans" with my helmet cradled like a baby in my arms, wondering why we weren't running out to the ramp to kick the tires and light the fires. Everyone would just sit around as if we had nowhere to go. It seemed like hours before my instructor would slowly get up and saunter toward the door, glancing at his watch, and suggest, "What do you say we go up two-ship with the Colonel, get the heck out of the coffin, and then maybe pick up a little contact over Hondo?" Calm as you please, as if we were going to wander over to the commissary for a jalapeño sub.

When Kerns at last realized how serious his situation was with his fatigued Air Force pilots, he recounted, "I fell to my knees and I prayed. Faith is a very important thing in your will to survive."

As Peter Leschak put it, "Whether a deity is actually listening or not, there is value in formally announcing your needs, desires, worries, sins, and goals in a focused, prayerful attitude. Only when you are aware can you take action." Survival psychologists have observed the same thing.

Kerns added, "All at once, it hit me that I might actually lose them. Those million-dollar pilots could die."

By chance, he found a fence and used the cedar post to start a fire. (Chance is nothing more than opportunity, and it is all around at every turn; the trick lies in recognizing it.) "It's amazing to see what fire can do. You're out in the woods, you're cold, you're lost, you're lonely. But the minute you light that fire, you're home, the lights are on, and supper's cooking. It made a world of difference going from complete darkness to light and warmth. It just turned everybody around."

Kerns learned many lessons that night. His mastery and confidence turned the pilots around even more than the fire. It showed them the way, and it made Kerns more able to save himself. That lesson was driven home again and again: Helping someone else is the best way to ensure your own survival. It takes you out of yourself. It helps you to rise above your fears. Now you're a rescuer, not a victim. And seeing how your leadership and skill buoy others up gives you more focus and energy to persevere. The cycle reinforces itself: You buoy them up, and their response buoys you up. Many

people who survive alone report that they were doing it for someone else (a wife, boyfriend, mother, son) back home. When
Antoine de Saint-Exupéry was lost in the Lybian Desert, it was the
thought of his wife's suffering that kept him going.

Mark Gamba is a steep creek boater and a photographer I've
worked with. When we were at Mount Hood together, he told me
of getting caught in a strainer, which is usually fatal. A strainer is
a tree that's fallen across a river. Ordinarily, people who are washed
into a strainer are pinned by the force of the current. Mark was
pinned, and each time he reached up to pull himself out over the
trunk of the tree, the current caught his legs and dragged him
back down. He managed to get his nose out of the water and
snatch a breath, but then he was pulled under again and again. He
was exhausting himself. His arms were burning. His lungs were
exploding. He began to see blackness creeping in from the edges
of his vision as the retina was robbed of oxygen.

"I thought of my son," he said. "I wanted to see him again."
And as he blacked out, he gave it one last try. In a tremendous
burst of energy, he pulled himself up and over the log.

In the south tower of the World Trade Center, an hour before
the building collapsed, Ronald DiFrancesco was one of the people
who met Brian Clark, the fire warden with the flashlight who was
asking people: "Up or down?" DiFrancesco went up, hoping to
find air. But after ten or so floors, he encountered people who were
succumbing to fatigue and smoke. The people, all of whom would
die, were just giving up and falling asleep. DiFrancesco, too, was
collapsing, but then he said to himself, "I've got to see my wife and
kids again." And with that, he got up and bolted down the stairs to
safety.

Doctors and nurses often survive better than others because
they have someone to help. They have a well-defined purpose.
Purpose is a big part of survival, but it must be accompanied by
work. Grace without good works is not salvation. The survivor
plans by setting small, manageable goals and then systematically

achieving them. Hence the Air Force checklist and the notion, which my father drilled into me: Plan the flight and fly the plan. But don't fall in love with the plan. Be open to a changing world and let go of the plan when necessary so that you can make a new plan. Then, as the world and the plan both go through their book of changes, you will always be ready to do the next right thing.

People are animals with animal instincts, but they lack many of the other survival mechanisms animals possess, such as fur to keep them warm, fangs and claws, and flight or speed. Culture creates a collective survival mechanism for the species. People survive better in numbers. They survive because they use cognition to organize, say, for a hunt, and to make things, even as cognition inhibits their animalness, including strength. That's why, when cognition is turned off, people are amazed by their own strength: because cognition continuously inhibits it. That is the whole secret to cognition: It is a mechanism for modulating emotional (physical) responses.

Every culture evolves survival rituals. Some, especially non-technical ones, are devoted to not much more than survival. In Native American cultures, one ritual is the vision quest, in which a young person goes into the wild and fasts in search of a vision. It can be seen as a type of survival training, for if there is no food, no water, no way, a person has already practiced sitting still and making the best of the situation. He'll have confidence in his ability to survive it.

The survival lessons that apply today are ancient. The *Tao Te Ching* is broken into two parts, "Integrity," and "The Way," which can be thought of as the two halves of surviving anything. Lao-tzu's book is a handbook for a ruler, but it is also a handbook for the brain. An imbalance of the brain's functions leads us into trouble, and a triumph of balance gets us out. I've found similar lessons in Epictetus, Herodotus, Thucydides, the Bible, the Bhagavad Gītā. "Is there any thing whereof it may be said, See, this is new?" says Ecclesiastes. But there are always new people who haven't heard that there's nothing new under the sun. And there's always some son of a bitch who doesn't get the word.

I had always wondered where our American survival rituals were. I think now that they're everywhere around us. The Boy Scouts in its original conception was a survival school. Sports are survival training in that they teach strength, agility, strategy, and the endurance of pain. But our culture is filled with survival stories as well. Cool is the ultimate American conception of the survival model. James Stockdale, a fighter pilot who was shot down over Vietnam in 1965, spoke many times about how he survived seven and a half years in prison camp. "One should include a course of familiarization with pain," he said.

Stockdale observed, "You have to practice hurting. There is no question about it. . . . You have to practice being hazed. You have to learn to take a bunch of junk and accept it with a sense of humor."

He's talking about being cool, just as he was when his F-8 fighter-bomber was hit with 57-millimeter fire. He had been on a relaxed bomb run at the time. He'd even taken off his uncomfortable oxygen mask. "I could barely keep that plane from flying into the ground while I got that damned oxygen mask to my mouth so I could tell my wingman that I was about to eject. What rotten luck. And on a milk run! My mind was clear, and I said to myself, 'five years.' I knew we were making a mess of the war in Southeast Asia, but I didn't think it would last longer than that." The Spartan practice of enduring the bite of the fox is "a course of familiarization with pain." Then there are the schools like Kerns's, which attempt to teach wilderness survival by directly meeting the problem head-on with Yankee ingenuity.

Like being lost, survival is a transformation; being a leader can ensure that, when you reach the final stage of that metamorphosis, it is with an attitude of commitment, not resignation. The transformation of survival is permanent. People who have had the experience often go on to become the best search and rescue professionals. They have come to understand, perhaps unconsciously, that they can only live fully by helping others through that same transformation. All the survivors I've talked to have told me how horrible the experience was. But they have also told me, often with deep puzzlement, how beautiful it was. They wouldn't trade the experience for anything in the world.

It gradually dawned on me that only by researching and dissecting the mysterious quality the Air Force so dully called Positive Mental Attitude would I ever understand survival.

And I thought: Wait a minute. My father was in the Army Air Corps. Maybe that's what he had that allowed him to live. If so, he'd certainly never talked about it. But what pilot would? I felt as if I'd stepped to the edge of the very thing I'd been after all my life. Here, concealed in the most unimaginative phrase possible, was the deep mystery I'd been trying to unravel.

THERE ARE two schools of thought on survival training. Byron Kerns's Mountain Shepherd Survival School is of the modern, technical sort. Mark Morey's Vermont Wilderness Survival School in Brattleboro, Vermont, is based on ancient native skills. Although they seem superficially different, I think they share important similarities.

When Jonas and I finished Kerns's course, we hopped into the plane and roared off, heading toward the Green Mountains. At Morey's school, we were led deep into the woods, where we found a group of children moving silently through the forest pretending to be deer. They were all about eight years old, and as we watched, they cupped their hands to their ears to amplify sounds in the quiet woods. An instructor snapped a twig, and they scattered into the forest and vanished. My first thought was that, if they tried that drill where I came from in Texas, they'd wind up riding home strapped to the hood of a Ford F-250 pickup truck.

We caught up with them 100 yards farther on, and Morey sat

them down and told them a story loosely based on Jack London's "To Build a Fire." As he finished and the main character froze to death, their eyes were wide. Then Morey said, "Now I've just fallen in a stream, and my feet are wet. I'm freezing cold. We have only one match. My feet are encased in ice. I'll die in five minutes unless I have a fire. Build me a fire. Go!"

The children began gathering tinder, kindling, and fuel, working quickly and without comment. As Morey counted off the minutes, they expertly laid the fire and had it going in four minutes, fifteen seconds. They worked harmoniously, putting self aside and helping each other. But the exercise wasn't entirely about firecraft.

"What we're doing," Morey explained as we made our way through the woods, "is loss-proofing these kids by building empathy and observational skills. In that last exercise, we demonstrated something we've discussed with them: The fight for survival requires a burst of energy. You have to become like a weasel, moving fast and getting the job done. Those kids just saved my life by building a fire. One day, they may save a life for real. They're learning to put the needs of the group ahead of their own."

Morey is one of the few people I've ever met who would survive being dropped naked into the woods. He's the Robinson Crusoe I was hoping to find when I reread Defoe's novel. He even looks a bit like the original cover illustration. (The real Crusoe was a bit disappointing: He had a Wal-Mart full of stuff to sustain him, which he salvaged from the grounded ship just offshore, including guns, powder, and livestock.)

It's fitting that Morey chose London's story. The unnamed main character in "To Build a Fire" was not aware of the forces he was entraining, though he'd been warned. He developed no secondary emotion to permit him to believe what he'd been told. He'd had just enough experience with the cold to make him think he already knew what was up. His cup was full, and he could not receive the wisdom of the old-timer who told him not to go off alone into the Alaskan wilderness. Motion held more allure than staying still, and so he engaged himself with an unstable system. The system was composed of his body, which had to remain at 98.6 degrees, more or less, and an environment where the temperature was 70 degrees below zero. With nearly 170 degrees of difference between the two, the second law of thermodynamics would ensure that any mistake on his part would send his energy level cascading rapidly down toward fatal hypothermia. He failed to anticipate the interaction of forces that would result from his stepping into a stream and flooding his boots. In moments, his feet froze solid. Now under stress, he failed to take in new information: In his haste, he built his fire beneath a tree that was laden with snow. It was one of those nonlinear kickback systems that was tightly coupled and that magnified trouble when upset. The Sand Pile Effect.

London described all this long before chaos or complexity theory had been developed. In a spare and systematic manner, his story touches on the most important elements of surviving:

> Each bough was fully freighted. Each time he had pulled a twig [from the tree] he had communicated a slight agitation to the tree—an imperceptible agitation, so far as he was concerned, but an agitation sufficient to bring about the disaster. High up in the tree one bough capsized its load of snow. This fell on the boughs beneath, capsizing them. The process continued, spreading out and involving the whole tree. It grew like an avalanche, and it descended without warning upon the man and the fire, and the fire was blotted out! The man was shocked. . . . Then he grew very calm.

The man had skills, equipment, and experience. It was his attitude that killed him, his inability to balance emotion and reason.

He wasn't a newcomer to that country. He'd been out when it was 50 below and he'd been fine. He'd gotten his feet wet before, built a fire, and been fine. But at 70 below, the system was driven so hard it exhibited a new and surprising behavior, one he'd never seen before. He had no idea of the transformations that can occur at such a critical boundary.

MOREY, JONAS, and I reached another group of children who were working on making fire with bow drills. I had never seen fire made that way and hardly believed it was possible. Then I watched a ten-year-old boy named Jacob make a roaring fire in about ten minutes. I asked him how long it had taken to learn. He told me two years. With a smirk, Morey asked Jacob if he was a master. Jacob said, "No. It could take a lifetime to master this." The skills were not mere skills. They were long-held practices that developed an attitude of quiet humility. They encouraged a slow, thoughtful, confident examination of the self and the environment.

That afternoon, I began work on my own fire, and it took me two grueling days. By the end of the first evening, I was frustrated, agitated, and irritated with this smug little Robinson Crusoe who could make flames leap from his fingertips on a whim. I didn't have any warm fire to sit by, and he wasn't going to make one for me. I'd had one MRE (Meal Ready to Eat) that day, and I was exhausted. All I had to show for my efforts was a bunch of strips of basswood bark, a few sticks, a chipped rock, and a ball of useless grass—it was making me angry. I wanted a flamethrower, a pizza, and some CDs. I wanted a drink.

It wasn't until I left the school and thought about it for a while that I realized fire was not the point. Morey was trying to teach me a path to seeing and knowing the world as well as myself. To find the materials; to know which ones would work and which wouldn't; to be on intimate terms with that new world. I could not change the world; I could only change myself. To see and know that world, then, was the key to surviving in it. I had to accept the world in which I found myself. I had to calm down and begin living. As with the Zen disciplines, the archery and martial arts, the practice of such skills could move perceptions and physical experience into a place where I could calmly take correct action. Whatever the neuroscientists eventually learn, one thing seemed certain: None of Morey's students is going to go out and ignore a big old rain cloud, as I had done in Glacier National Park. Each of them is going to smell it, see it, feel it, long before it comes. He's going to ask himself: What is the correct feeling of this? What is the correct action? Morey's training was making it impossible for them to embark on the stages of becoming lost, of dying. His training would keep their mental maps true to the world, keep emotion and reason balanced, keep the eyes and ears open.

Living isolated from nature, I'd gone to sleep. Safety, convenience, and efficiency are the opiates of the modern world. But I began to see that there were kinds of training that could wake me up. When I was first learning aerobatics, I was scared to death most of the time. I asked my father how he'd gotten comfortable going on those bomb runs, thinking that I might transfer the learning to the risks I was taking. "Don't get comfortable," he advised. "Get confident."

As we walked through the woods, Mark would stop me and Jonas from time to time and make us listen to the birds. "Something's up," he'd say, his whole body as alert as that of a hunting dog, while his eyes darted this way and that. "Listen to how excited they are. The birds are the radar of the forest." I recalled a day when I was camping in Big Bend National Park on the Texas-Mexico border. I'd gone up into the Chisos Mountains with a naturalist named Anita to escape the heat. The previous night, the thermometer had read 108° at nine o'clock. I was in that frame of mind where bad things happen: tired, dehydrated, and with a fair case of get-there-itis. But the heart-wrenching cries of a blue jay caught Anita's attention. She put out her hand and stopped me on the trail.

I looked up and there she was, a blue jay in a small tree by the trail, just wailing her heart out. For what? Anita pointed at the trail. There, lying across our path, was a big rattler. I'm sure I would have stepped right on it if I'd been alone. Anita was attuned to her environment, alert to asking: What's up? The snake had just eaten the blue jay's baby.

At Mark Morey's school, I eventually managed to make fire with a bow drill. I can't take the credit. It was firemaking with training wheels. Morey helped me every step of the way—gentle, prodding, encouraging. Also, at the end my spindle slipped, sending the fragile coal flying. But Morey delicately picked it out of the grass with a dry leaf and transferred it to a tinder bundle he'd prepared for me. (Making a tinder bundle is an art in itself—it can take years to get right.) But when I cupped it in my hands and blew it to life, its power was immediate and magnificent. I felt as if I'd watched a baby being born.

Like making art, making fire is a deeply human act. Through it, we know our world in a way that no animal ever will. I felt like a sorcerer. The burning bush had spoken.

MOREY'S SCHOOL is full of drama, passion, and surprises. Its lessons are meant to extend the deep connection between awareness and survival, to let us think with the mind of a child, the Zen mind, the beginner's mind. When he told us we could navigate without compass or map, Jonas and I both were pretty skeptical. We'd just come from an Air Force–style survival school where a knowledge of map and compass was paramount. But the lesson was not about finding our way in the woods; it was about navigating the human brain.

As we hiked through trailless forest, Morey stopped every 20 or 30 yards to point out something, and we'd examine and discuss what we found. After we'd followed him deep into the woods, he asked us to close our eyes and point the way home. It is a humbling experience to find that you can't. I'd been following him, which is never a good idea. I had not walked my own walk, and as a result, I was lost.

Morey directed our attention to the last place we'd stopped to talk. We could still see it from where we stood. "Remember, we talked about the bittersweet vine there?" We'd taken a sample from a vine that's good for making cordage. So we hiked back to that spot. Then he pointed to another spot, where he'd shown me ways of seeing and walking that were used by Native American trackers and other Aboriginal peoples. He called it "Owl Eyes and the Fox Walk," that full-body alertness I'd seen when he listened to the birds. It can put you in an altered state of perception, he said. We returned to that spot. From there, we could see the place where we thought we'd found the hoof print of a deer, but it turned out to be the entrance to a vole tunnel. We had squatted there to discuss the difference between voles, moles, and mice.

Thus, hopping from one conversation to the next, we were able to retrace our steps exactly and to remember in great detail not only where we'd been but what we'd said and done at each spot. In what seemed to be a featureless and homogenous forest, Morey had given us tangible cues, like road signs, which we could easily follow home. He had discovered an effortless way to embed a reliable mental map in our brains.

"It's called song lines," he said. "And it's an ancient navigational technique used by Australian Aboriginals."

In Australia the Aborigines have a "labyrinth of invisible pathways," as Bruce Chatwin wrote in *The Songlines*, "which meander all over Australia and are known to Europeans as 'Dreaming-tracks,' or 'Songlines'; to the Aboriginals as the 'Footprints of the Ancestors' or the 'Way of the Law.'

"Aboriginal Creation myths tell of the legendary totemic beings who had wandered over the continent in the Dreamtime, singing out the name of everything that crossed their path—birds, animals, plants, rocks, waterholes—and so singing the world into existence. . . . A song . . . was both a map and a direction-finder. Provided you knew the song, you could always find your way across country."

"If you do that when you go into the woods," Morey told us, "you'll never be lost."

I was cynical at first about pretending to be a deer or an owl or a fox. But when I phoned him weeks later, Morey was able to recite in detail the steps in our journey. He remembered specific fallen trees, a patch of Japanese knotweed, and the way the wind was moving the leaves on a stand of maples. "It's not like I tried to remember that," he said. "It's an ancient instinct. And it's still alive." It was the first time I'd heard a strategy for making the mental map match the world. Map and compass are artificial methods for doing that, and they work well. But this was . . . deep. It was aboriginal neuroscience, using implicit, not explicit, memory circuits to embed the map in the unconscious mind.

Each child in Morey's school selects a "secret spot" in the woods. Every day, the students are directed to spend some time at their secret spot. "They do this through all the seasons," Morey said. "They learn to stop and think. They learn to be calm and alone. They will never feel that the woods are alien. So, if they ever get lost or otherwise stranded, they won't panic." People revert to automatic behavior, implicit learning, in an emergency. These children will automatically do the right thing. They will never lose their way.

I WENT home enthusiastic for Morey's way of walking in the wilderness, but I quickly realized it would have to be a lifelong commitment. He estimated it would take a minimum of five years to learn those skills well enough to depend on them in the woods, and that would be only the beginning.

On the other hand, even those of us who can't make such a commitment can benefit from his lessons. Ken Hill, for example, told me that you can learn to pay a different kind of attention in the woods without becoming a fanatic. You can learn map and com-

pass work in a few weeks of practice. You can also learn to turn
around and see where you've been, which is a means of paying
attention. If you come to a fork in the road, you look back and find
a cue, something different, and remember it. Perhaps talk about it.
It's a modified songline. You don't have to be perfect; you can simply be better.

ELEVEN

"WE'RE ALL GOING TO FUCKIN' DIE!"

IN JANUARY 1982, Steven Callahan set out to cross the Atlantic alone in his boat, *Napoleon Solo*. He started out in the Canary Islands after sailing from Lisbon. On January 29, having determined from the charts that there was a 2 percent chance of a gale at that time of year, he steered away from the island of Hierro, estimating his arrival in the Caribbean by February 24. With a video camera running to document the voyage, he roared into the teeth of a gale on February 4.

Callahan had designed his boat well, though, and he wasn't concerned. He had a glassed-in pilothouse from which he could steer, and plenty of experience with storms. It was two days before his birthday. Just before midnight, he ate some chocolate, did an inspection, and went to bed wearing nothing but a T-shirt. It wasn't the weather that got him. He'd done everything right. His was a "not your day" kind of accident.

He believes that he may have been struck by a whale. Whatever the cause, he shot out of bed at the sound of a loud noise and a rush of water exploding into the cabin. Before he could even get to his knife, he was waist-deep in water and had to dive in total darkness to cut the lines on his survival bag. He failed.

As it was, Callahan barely got out with his life. Awash in water on deck, he managed to get his life raft out and into the heaving waves. He put the knife in his teeth before abandoning ship. Then he noticed the video camera. "Its red eye winks at me," he wrote in his book, *Adrift*. "Who is directing this film? He isn't much on lighting but his flair for the dramatic is impressive." Even as his entire life-support system was being wrenched out from under him, Callahan was able to stand outside the situation and laugh at himself, pretending to be the dashing buccaneer. That's some admirable cool.

Starting from the moment of the accident, it is necessary for a survivor to take control of his situation. Peter Leschak speaks of "Standard Fire Order #10: Stay alert, keep calm, think clearly, act decisively." According to this directive, the best way to meet an emergency is with sharp senses (to gather information), a clear mind (to analyze the information), and bold action; add to these humor (to handle strong emotions). Steven Callahan was able to do all of those things.

Compare Debbie Kiley's experience just three months earlier, when the ketch rig *Trashman* was sinking on the opposite side of the Atlantic. She and a small crew had set out on a routine yacht delivery when they encountered a hurricane. Mark Adams, the first mate, was screaming, "We're all going to fuckin' die! We're all going to fuckin' die!" In his panic, he inflated the life raft before attaching it to *Trashman*. The wind immediately picked it up. "It blew through the rigging, skimmed the top of a wave, and vanished," Kiley reported in her book (written with Meg Noonan), *Untamed Seas*. Not cool.

John Leach puts it this way: "When the personality is ripped away there has to be a core remaining to carry the person through. . . . If a person can carry all his support within him then it matters little what the external environment comprises." Adams had no core. Callahan did.

Survival starts before the accident. *Napoleon Solo* didn't sink immediately but floated for a while. That was a result of Callahan's meticulous planning: Callahan, a boat designer, built the 21-foot cruiser himself and had included in its design several watertight compartments to prevent or prolong sinking. He kept his raft tethered to *Solo*, watching to see how the situation would develop. Just for the experience, he had once inflated the four-man life raft that the U.S. Coast Guard required on his boat. He had climbed inside with two friends and realized how grim it would be, trying to drift at sea in such a torture chamber. So he equipped *Solo* with a six-man raft. The life raft had its own survival supplies, of course, but he'd provided more. Unfortunately, he'd been unable to free the large, well-stocked survival bag he'd prepared. Now, as he waited, considering his next action, he looked up and saw the moon and suddenly noticed how sharp his senses had become, an early sign of survival behavior. His perceptions didn't narrow with fear, they opened up. This phenomenon has been demonstrated experimentally. LeDoux points out that certain people, when afraid, experience "activation of the amygdala [which] will lead working memory to receive a greater number of inputs, and inputs of a greater variety, than in the presence of emotionally neutral stimuli."

Callahan was fighting a battle for control, though, and the outcome was far from decided. Cool cultural icons don't usually tell the other side of the story. It's not that cool people are unafraid. They're very afraid. But they are, as Tom Wolfe recorded, paraphrasing Saint-Exupéry, "Afraid to panic." This is how Callahan described those first few minutes in his 1986 best-seller *Adrift*:

Myriad conversations and debates flash through my mind, as if a group of men are chattering within my skull. Some of them joke, finding comic relief in the camera's busily taking pictures that no one will ever see. Others stoke a furnace of fear. . . . I must be careful. I fight blind panic: I do not want the power from my pumping adrenaline to lead to confused and counter-productive activity. I fight the urge to fall into catatonic hysteria: I do not want to sit frozen in fear until the end comes. Focus, I tell myself. Focus and get moving.

Turning fear into focus is the first act of a survivor. But Callahan's experience of myriad emotions at the same time is also a common reaction reported by people caught in the midst of a catastrophe. When the World Trade Center was hit, Sam Melisi, a firefighter on a fireboat nearby, reported that he "had a strange impression of feeling every possible emotion all at once." Saint-Exupéry observed that, in the heat of a crisis, the only thought you can allow yourself concerns your next correct action. In describing his passion for the lethal profession of firefighting, Leschak wrote of "the dreadful/wonderful moments when fear makes you so alive you simply cannot die."

At least 75 percent of people caught in a catastrophe either freeze or simply wander in a daze, according to some psychologists. They can't think, they can't act correctly. (Others maintain that people rarely panic, but often remain calm in denial, as happened in the World Trade Center.) When the United States attacked Hiroshima and Nagasaki with atomic bombs, the Japanese noticed the same phenomenon among their survivors. They named it *burabura*, which means "do-nothing sickness." In the days following the collapse of the World Trade Center, I noticed it among the people who kept returning there.

But already, Callahan was formulating a plan for action. He knew that he should stay attached to the boat as long as it was floating. He knew that he wasn't in the shipping lanes. His chances of being picked up were poor. The next step was to take bold action while exercising great caution, which is but one of the many delicate balancing acts necessary for survival. So he dove back into the flooded saloon to retrieve his survival bag. He made it out and returned to the raft. When he eventually let go of *Solo*, he was very well equipped, considering the circumstances. He had just saved his life by risking it, which is the essential task of every organism. No risk, no reward. No risk, no life.

All the time, Callahan was analyzing his situation and working to keep his emotions with him, not against him. He knew the current would carry him west, not east, and that the first landfall was in the Caribbean, 1,800 miles away. He knew that to think of the impossibility of drifting there would excite dangerous emotions. Instead, he planned only as far as the following morning. By not thinking of the almost certain death that would result from trying to drift 1,800 miles, he was keeping himself from despair and panic. So he set up a small, attainable goal. To act is human, to succeed, crucial. He was giving himself something to do and a chance for success, both of which are necessary early in the survival scenario.

Again, Callahan found humor, chiding himself, "in a Humphrey Bogart fashion. Well, you're on your own now, kid." It was no coincidence that he thought of Hollywood movies, because they provide us with our cultural archetype for cool, from Brando and James Dean to Bogart and Dashiell Hammett's Continental Op, who believed, according to Claudia Roth Pierpont, writing for *The New Yorker*, "emotions are a nuisance during business hours, and all his hours are business hours." Every culture has its survival rituals, and ours are found in Hollywood movies and pulp fiction. The hero stays cool while the villain is ruled by emotion. That's why he dies.

A sense of humor "is not a luxury," Leach writes, "it is a vital organ for survival." Foxhole humor is well known among soldiers and is an essential ingredient for survival anywhere, from being adrift at sea to finding yourself in the middle of a divorce or enduring a loved one's death.

But emotion has power, too. The trick is to harness that power in the service of survival. "Fear becomes sustenance," Callahan wrote. "Its energy feeds action." That's what got him down into the dark, flooded cabin of the boat. It's also what got him out.

Another factor in Callahan's survival was how generally well prepared he was. In addition to all the practice he'd had, all the experience at sea, he had read a great deal. He even carried Dougal Robertson's book *Survive the Savage Sea*, an account of being sunk by a whale and drifting for thirty-eight days. He'd read about the Baileys, Maurice and Maralyn, who described in *117 Days Adrift* how they'd survived that long at sea. Callahan knew that few castaways made it past a month; but significantly, he knew that it was possible. He knew something that every survivor must bind to his heart with hoops of steel: *Anything is possible.*

Callahan began solidifying his resolve. "I've got to do the best I can," he told himself. "The very best. I cannot shirk or procrastinate. I cannot withdraw . . . I have sometimes fooled other people. But Nature is not such a dolt." He had adopted the attitude of humility so important to survival. A Navy SEAL commander told Al Siebert, the psychologist, "The Rambo types are the first to go." Scott observed the same thing during his explorations in Antarctica. The concept is ancient. The *Tao Te Ching* says:

> He who is brave in daring will be killed,
> He who is brave in not daring will survive.

So far, Callahan had acted like a perfect textbook survivor. He was consolidating his personality and fixing his resolve. He was also, again, putting to use the lessons in the mythology of the culture, all those survival myths that he was raised on. He was using those images to form a memory of his possible future, making it just as real as his past.

WHEN YOU are adrift at sea in a survival raft, within twenty-four hours your skin succumbs to hundreds of sores caused by the salt water. Callahan couldn't sleep because of the pain. And yet, when he managed to patch a small hole (already!) in the floor of the raft, he felt victorious. To take delight in small achievements, to celebrate early victories, is another hallmark of the survivor. When his spirits lifted, he remarked, "I have risen, if briefly, from my death bed."

Though it was only the first full day, Callahan had already completely converted himself from victim to survivor. It's unusual to find someone engaged in recreational activities who can achieve this so quickly and seamlessly, and with such style and grace. Those who do so are usually highly trained professionals—fighter pilots, elite mountain climbers, or firefighters.

Compare Callahan's reaction with that of another sailor, Tami Oldham Ashcraft. Her boat lost its mast in a hurricane. Although she didn't lose her vessel, her circumstances were certainly dire. Her fiancé died in the storm when he was thrown overboard, and that has to be taken into account. And now, with no engine or radio, she couldn't expect help. But she had food, water, gear, and was able to rig a sail to move the boat along and take sightings with her sextant (Callahan had to rig one up using a pencil). Ashcraft managed to sail to within sight of land, but then it was obscured again briefly. "I collapsed onto the seat locker," she reported later in *National Geographic Adventure* magazine. "'This is hell. I'm in hell . . .' Hysterical, I dashed down the companionway and grabbed the rifle and a box of cartridges from the locker. I loaded the rifle and, leaning against the nav station, tried to cram the barrel into my mouth."

On September 11, 2001, Pasquale Buzzelli, an engineer working in the north tower of the World Trade Center, was in an elevator at the moment of impact. He got out at the forty-fourth floor, and despite pandemonium and heavy smoke, stepped into another elevator to his office on the sixty-fourth floor. He was persisting in denial, an emotional state that can have far-reaching effects. Everyone on his floor was calmly waiting for security to tell them what to do. He proceeded to call his pregnant wife and reassure her that everything was fine, even though she could see on television that an airliner had crashed into her husband's building.

Carrying on in a kind of dream state, Buzzelli and his colleagues simply hung around until the south tower collapsed. Not even that event motivated Buzzelli to leave. What convinced him at last was the thickening smoke. He and the others started down the stairs, and Buzzelli was between the twenty-first and twenty-second floors when the tower collapsed. He was conscious the whole time, praying that death would take him without long suffering. He remembers the fall vividly, the rhythmic thunder as the floors pancaked one on top of another, and the rocket rush of the ride down. At last he was knocked out and remained unconscious for three hours.

When Buzzelli awoke, he was outside in the open but trapped high atop a piece of the wreckage, unable to get down, with fires burning nearby. When firefighters eventually spotted him, they were unable to reach him. They needed to round up some equipment before they could get him down. After they left him alone, Buzzelli panicked about being killed by the fire, even though he couldn't see any flames. So he decided to kill himself. He picked up a sharp fragment and was preparing to cut his own wrists when the fire went out of its own accord. When the firemen returned, Buzzelli changed his plan.

Tami Ashcraft didn't shoot herself, either, and both she and Buzzelli were rescued, but their attitudes are not the survivor's way of looking at the world. Curiously, nonsurvivors don't always perish and true survivors sometimes find themselves in circumstances where the objective hazard is simply too great, as it was for many of the victims in the World Trade Center.

DEBBIE KILEY'S and Steve Callahan's experiences present a dramatic contrast in both behavior and outcome, despite the similarity of their accidents and circumstances. This is not meant to be a criticism of Kiley, a brave and excellent sailor. It is simply a comparison for the sake of analysis. Given the crew and supplies she had, the objective hazard she faced was overwhelming.

John Lippoth, captain of the *Trashman*, and Mark Adams, his first mate, were heavy drinkers. In addition, Mark was given to fits of very strange behavior, ranting, jerking women's panties down in bars, and veering toward violence. Brad Cavanagh, another member of the crew, was an able seaman and not a drinker. Meg Mooney was Lippoth's girlfriend and had never been on a voyage on the open ocean before. Kiley had sailed professionally; indeed, she was the first woman ever to sail the Whitbread Around the World race.

Trashman's trip was intended to be a simple yacht delivery from Bar Harbor, Maine, to Fort Lauderdale, Florida. In the 58-foot ketch, it should have been routine. Despite her bad feeling that she should not go with that motley crew, Kiley allowed herself to be bullied into it by Lippoth.

But once at sea, she realized what a mistake it had been. Lippoth and Mark Adams were drunk the whole time; Meg was useless; and she and Cavanagh were left with all the work and none of the sleep. Then they sailed into a hurricane. After one of her few breaks from being at the helm, Kiley came up on deck to find Mark raving drunk and "howling into the [55-knot] wind." The seas were "walls of liquid granite 30 or 35 feet high." Kiley and Cavanagh took over, leaving Lippoth and Adams to drink down below. When they were too tired to go on, they demanded that Lippoth and Adams take their turn. Kiley had to sleep. She was awakened a short while later in the middle of the afternoon to the news that the boat was sinking.

All that was left after Adams lost the raft was an 11-foot Zodiac, an inflatable runabout with no supplies in it. As they were trying to get the small, semirigid boat into the water, Meg was thrown into the *Trashman*'s rigging by a wave and gravely injured.

Even as the crew swam from the sinking yacht, with Adams screaming hysterically, the Zodiac flipped upside down, and the five people were left clinging to the line that ran around the outside of its hull. Within two minutes, *Trashman*'s masts disappeared beneath the waves. In those first minutes, Kiley was fighting her panic and experiencing that rapid shift from denial to extremely acute sensory input.

"Am I really here?" she asked herself. "I could smell rubber and taste salt on my lips. I became keenly aware of the warmth of the water." She struggled for some sort of control but found herself overwhelmed by the chemicals pouring into her system. She immediately began trying to prepare for survival, but the others were lurching haphazardly from hysteria to shock. Her situation was more difficult than Callahan's in that she had to fight not only the sea and herself but the crew. Given the way she tried to behave, she probably would have done much better either alone or with no one but Cavanagh.

The wind and waves kept peeling their hands away from the Zodiac's line, and when they managed to get the boat upright, the wind would flip it again, tossing everyone out. Nature loves to strip the unwary of their gear. Meg had been stripped of her clothes; even her underpants were gone. Only her shirt remained. At one point, while trying to help Meg into the boat, Kiley noticed how badly injured the young woman was, her legs severely lacerated from the rigging cables. Kiley finally got the others to ride out the storm by sheltering beneath the inverted dinghy, where they had at least some protection from the wind.

As night fell, Kiley groped for any useful thoughts, but her struggle with her emotional reactions was not going well. There had been little in her life to prepare her for such an event. Her racing experience had been with a competent, well-equipped, professional crew, and she had been able to depend on them if anything happened. Now she was adrift with no food, no water, no clothing, no signaling devices, nothing but what they were wearing.

The closest Kiley came to survival thinking that first night was when a ship passed near them. While the others were cheering wildly, "the reality of our situation came back into focus," Kiley wrote later. She knew that the chances of their being spotted, even by the Coast Guard, were low at night. She understood that they were on their own. She was thinking analytically, assessing her situation, and accepting reality. She was able to face the notion that she would have to take responsibility for her own survival.

At one point, Mark complained that Kiley was kicking him, and knowing that she wasn't, she ducked underwater to see what he was feeling. "A cold sword of fear stabbed through me. I didn't believe what I was seeing. . . . Sharks. There were sharks everywhere." There could not have been a better lure for sharks than the inverted boat with all those legs dangling down and trailing Meg's blood. Soon the boat was upright and everyone was inside it. But now, exposed to the wind, they were in danger of hypothermia.

After that, they were harried by sharks on into the next day. They nearly flipped the boat when, fearing a shark attack, everyone moved to the same side at once. There was no discipline, no leadership, no plan, and a lot of screaming and fighting. Emotion was a loose and loaded cannon. Kiley was not immune to it, either, but she kept trying to direct her thoughts and control her emotion. Her success was limited at first, but at least she was engaged in a struggle. Without the effort of struggle, a person is done for. Just as James Stockdale recommends "a course of familiarization with pain," and Mark Morey observed that children without knowledge of predators don't do as well in survival school, so people who have not had to struggle in their lives are at risk.

While the other crewmembers veered wildly from hysteria to giddiness to deep depression and angry, irrational arguing among themselves, Kiley fought to consolidate something inside herself. "I closed my eyes, trying to get away from their death masks, from the sharks, from everything." She was looking within herself for balance.

The balance was there, and the evidence is in her state of mind in the moments after impact, which resembled Callahan's in the first few minutes after *Solo* sank. "My mind seemed trapped in this rapid-fire volley," Kiley reported, "thoughts rocketing by like snipers' bullets. Control, I said to myself. I must maintain control." She would, but it would take work and time. Even so, as if by some long-forgotten schooling, she began doing some of the right things almost immediately. Struggling to achieve that essential state of grace and poise, she began praying to keep herself focused. Survival psychologists have long observed that successful survivors pray, even when they don't believe in a god.

Kiley tried to organize the crew to scoop some warm seawater into the dinghy to protect them from the cold wind, but her efforts were fruitless. When Mark Adams thought he saw their lost life raft, everyone flew into a frenzy of paddling with their hands, wasting precious energy.

At last, Kiley hit on the idea that they could cover themselves with seaweed for warmth. An inventive approach to using the materials at hand is a hallmark of survivor thinking. Although it was a good idea and eventually worked, Mark's initial response was, "I don't want that slimy shit on me." You can often tell early on who is going to make it and who is not. If Kiley's story had been a Hollywood movie, everyone in the audience would know by now that Mark was going to die.

Meanwhile, Brad Cavanagh was sitting alone by himself on the bow of the boat, shouting at the sky: "Fuck you, God. Fuck you, you fucking bastard!" He was doing the opposite of going inside himself, where survival begins. Epictetus wrote: "You must either cultivate your own ruling faculty, or external things; you must either exercise your skill on internal things or on external things; that is, you must either maintain the position of a philosopher or that of a common person." And: "The condition and characteristic of a philosopher is this: he expects all advantage and all harm from himself." He doesn't blame others, nor turn to them. He takes responsibility for himself. Epictetus, one of the great Stoic thinkers, was writing a survival manual. And what he said was this: If you want to be a survivor, be a philosopher.

Cavanagh blamed something outside himself for his predicament. He was not looking within for consolation and consolidation, for opportunity and strength. All five crewmembers were, quite literally, dying. They were progressing through Kübler-Ross's stages: denial, anger, bargaining, depression, and acceptance. But although it is instructive to compare their behavior to Callahan's (who was not dying but surviving), it mattered little what they did. Through lack of proper preparation, they had no water. Without rescue, they would perish.

It's not uncommon. On the contrary, as John Leach has observed, "People will quite simply ignore the possibility that a particular disaster will ever strike them." Some people subconsciously believe that "to prepare for disaster is to encourage it. 'Don't even think about it'—for fear that it may come to pass."

Kiley continued intermittently to grasp at her inner instinct for survival, but she was continually pulled back toward thinking and behavior that could be deadly. She entered into a phase of bargaining, telling God, "If you get me through this I'll try to do better." Many victims have perished waiting for God to help them out instead of recognizing that, whether or not you believe in a god, you must help yourself.

Still, she was doing her best, struggling to think like a survivor. When she found that the seaweed with which they'd covered themselves sustained a vast number of tiny creatures, "I was dazzled by the life it supported . . . an entire world, self-sufficient and complete." To be open to the world in which you find yourself, to be able to experience wonder at its magnificence, is to begin to admit its reality and adapt to it. Be here now. It is to place yourself in relation to it, to say: Before I came here, the world was as it is now; after I am gone, it will be that way still. To experience wonder is to know this truth: The world won't adapt to me. I must adapt to it. To experience humility is the true survivor's correct response to catastrophe. A survival emergency is a Rorschach test. It will quickly tell you who you are.

What Debbie Kiley perceived was this: "I was frightened by the way I seemed to be losing control of my thoughts." She understood at a deep level the need to gain the upper hand, the high ground of the forebrain.

Meanwhile, Mark Adams was still screaming. "Didn't anyone tell them we were fucking out here!" he raved. He had no idea who might have told "them" or who "they" were. He didn't plan for the situation because he didn't know his own world or take responsibility for himself. He expected the world to adapt to him, to take care of him. At the same time, John Lippoth had fallen into a stupor, and Meg was succumbing to her injuries.

As darkness fell again, Kiley let her thoughts take her to another world, a refuge that survivors seek, which can help them to regroup and gain energy: long-term memory. "There was a ball game in a mowed green field and children on backyard swings and dogs barking and cars turning into driveways," she recalled. The struggle to survive can rapidly deplete your resources. Balancing emotion and reason can be like teetering on top of a big ball, requiring touch, timing, and a continuous shift of weight and energy. (The phenomenon may have something to do with stimulation of the amygdala, too. Ian Glynn, author of *An Anatomy of Thought*, wrote that "hallucinations in which the patient relives past experiences may be elicited . . . by stimulation of the amygdala." Hallucinations are frequently reported by survivors.) That retreat into memory, fantasy, or mind games can bring relief and rest. It may also help that the brain can do its work on something familiar when all the normal cues are stripped away. In the brain's desperate struggle to place you back in the world, it can bring out memories of hospitable environments the organism can tolerate. The struggle to adapt to such a hostile world can be exhausting, and memory of a benign environment can provide a necessary refuge.

Henri Charrière, the French writer and convict better known as Papillon, was condemned to life in the infamous penal colony on Devil's Island. He wrote: "Once I was truly exhausted, I stretched out on my back and wrapped the blanket around my head . . . what indescribable sensations! I spent nights of love, more intense than the real ones. I could sit down with my mother, dead these 17 years . . . I was there; it wasn't imagination." He escaped from the island, first in his mind and eventually in real life.

But without training or knowledge of survival, without something in her background to anchor her, Debbie Kiley was unable to grasp what was happening to her. Still, she was in better shape than the others. She felt empathy for Meg, for example. She knew that surely Meg would die and "wondered if she was thinking about home right now." Empathy is an important quality for a survivor. Kiley's desire to aid Meg may have helped her feel less like a victim. It may have given her at least some sense of personal power.

Whereas Steve Callahan had begun his solo voyage with the understanding that he had to take care of himself, in Kiley's group no one had worked out beforehand who was in charge, who was responsible; certainly not the captain, Lippoth. Routine, rules, emergency procedures were all left to chance. Now they were all thinking outside themselves, hoping that some mysterious force would come down and snatch them out of their nightmare. The failure to adapt was killing them.

After a time, Brad Cavanagh began making an effort to rise above the group and his own emotions. He had gotten up on the windward side of the gunwale to prevent the boat from flipping. He'd been very quiet. "Brad sat stone-faced and silent up on the side of the Zodiac. He had his mission; he was the balancer of the boat." He was separating himself from the group's poisonous dynamics and consolidating something within himself to resist their descent into madness. Significantly, he did it by giving himself a job, a purpose, a small task.

The situation was very bad. They were hurt, exhausted, hypothermic, and had no water. Two days out, their pain was scarcely imaginable, and some were less than twenty-four hours from death. More and more often now, people in the group were saying, "I can't take it." But that is the exact opposite of what a survivor would say to himself.

Christopher Burney, the British officer who was held in Buchenwald and other German prison camps, was kept in solitary confinement for years during World War II. At first, he told himself he'd be out by Christmas. When Christmas passed, he hoped to be released by Easter. When that, too, passed and summer came, "I dismissed my old impatience from my mind," he wrote in *Solitary Confinement*, "seeing such promise in the summer weather that no reservation, with its hidden pessimism, was now necessary . . . I could be patient for three more months." That is the way a survivor thinks. When I was working in maximum-security prisons in the early 1980s, I remember one convict telling me, "I could do a nickel standing on my head." When I asked how he did it, he said, "You got to stay inside yo' mine." That's survivor thinking.

Yossi Ghinsberg, a young Israeli hiker who was traveling on a raft downriver and then got swept into rapids in the Bolivian jungle, stumbled into a pit of quicksand during his third week of peregrinations. As he lay gasping for breath, after having struggled out, he told himself, "I'll suffer any torment, but I'll go on." There are few perfect survivors. Steve Callahan may be one. There are many, though, who flail around at first, then get their minds right, and live.

AS THE group deteriorated even further, as it became obvious that Meg was dying, Kiley fought harder for control. She was not going to give up, and that marked her, at last, as a survivor. At one point she covered her head with a rubber spray curtain near the boat's bow "to separate myself from them . . . to keep myself together. They were dragging me down, filling my head with black thoughts." She attempted to forge a team with Brad Cavanagh, the only other crewmember who wasn't mad by now.

"Brad," she said. "I need you."

He put his head under the apron with her, and they made a pact. One of them would sleep while the other stood guard. It was simple, but for the first time, it began to define some measure of control and to separate the survivors from the victims. It gave them a simple task, a short-term goal, a purpose, and a commitment to someone outside themselves. It also solidified enough coldness in their hearts to let go of the other three crewmembers, to do . . . *whatever was necessary*. Their empathy could now be directed toward each other and no one else. Like an immune system, it defined the inside and the outside. And by being responsible to each other, neither had to feel like the victim.

Years later, when I asked Kiley how she and Brad Cavanagh had survived, she told me: "I got Brad on the boat and I felt like it was my job to keep him alive. And I think he thought the same about me. Brad and I had a little rule: If I'm asleep, you're awake, and if you're asleep, then I'm awake."

With that, Kiley at last began consolidating what she had, examining her resources, mapping her world and adapting to it. She remembered her high school English teacher, Mr. Pitman. He had taken the class on wilderness trips and talked about survival. "I hadn't thought about him in years." Perhaps there was something in her background she could use. "No one is coming," she told herself. "I had to face facts. We all did. We were on our own."

By then, Adams had gone completely simple. Seeing a cut on Kiley's foot, he tried to twist her toe off, telling her, "It's so infected it's just going to fall off anyway." Kiley was unmoved. "I looked away. The sight of them made me sick." To survive, you must at some point allow cool to become cold. Stockdale wrote, "In difficult situations, the leader with the heart, not the soft heart, not the bleeding heart, but the Old Testament heart, the hard heart, comes into his own." Survival means accepting reality, and accepting reality takes a hard heart. But it is a strange kind of coldness, for it has empathy at its center.

Survivors discover a deep spiritual relationship to the world. They often have a talisman to connect them with it, a sort of lifeline from inner to outer reality. Debbie Kiley found her spiritual side and her talisman in the memory of an old sailor who had told her that "if a sailor wearing a black pearl was ever lost at sea, he could trade it to Poseidon for his life." After hearing that, she had bought a black pearl earring, which she still wore. "Now I reached up and touched the earring. I fingered the pearl." She took it off and threw it into the sea. When Brad asked what she was doing, she said, "Nothing." She recognized, too, her need for privacy: to keep certain things for yourself and maintain that consolidated identity, that sacred inner core of selfhood. Orwell's novel *1984* is in part about how Winston Smith gradually loses that sacred core. In group survival, we all need privacy. The immune system itself tells us what is us and what is the rest of the world, as do emotions. Privacy is essential to life. Life itself can be seen as arising from a self-organizing force that gathers certain materials, hoards and tends them, and protects them from becoming part of the rest of the world, even while delicately interacting with, finding a place in, taking from and giving to the world. Privacy is life, but so is community. It's another balancing act, to have boundaries and not be completely alone. At times when nature is trying to reclaim the materials of life, to turn you into the raw stuff of the world for its eternal tinkering, you have to cling to what is yours and yours alone, even while committing to another. That may be the meaning of a talisman: This is me, this is my stuff. Yossi Ghinsberg clung to a little book that his uncle had given him. Steve Callahan clung to his Boy Scout knife. Kiley clung to her black pearl and committed it to Poseidon. She made a pact between herself and the sea.

Although Kiley was beginning to find a calm within herself, she still had at best a tenuous hold on it, and at times, found herself asking one of the deadliest questions of all "Why this? Why now? Why me?" Gratitude, humility, wonder, imagination, and cold, logical determination: those are the survivor's tools of mind.

KILEY WAS asleep when Brad woke her urgently on the morning of the third day. They watched as John and Mark leaned over the side and drank their fill of seawater. "I felt such sadness listening to them," Kiley wrote. "They had given in, they had lost control. I knew that drinking seawater was a terrible mistake."

Even as they gave up their struggle, Kiley seemed to strengthen. She began to experience the all-important split of personality, that paradox of hope and resignation, which characterizes survivors. "Maybe the Coast Guard was coming after all," she told herself, answering, "Don't be a fool, Debbie. . . . They aren't coming; that's all there is to it . . . I will help myself, I thought. All I have is myself. I will not fall apart."

By late afternoon, as Lippoth, Meg, and Adams all indulged in complaint and bickering, Adams suddenly began searching the seaweed in the bottom of the boat. "Where are my fucking cigarettes?" he asked. When Kiley told him there were no cigarettes, he grew angry. "I just went to the Seven-Eleven and I bought beer and cigarettes. And I want to know who took them." Meanwhile, Lippoth began deliberately kicking Meg, who was already in terrible pain from her wounds. As Adams continued to search, Lippoth began peeling away a patch on the inflated tubing that was keeping them afloat. "My knees buckled," Kiley wrote. "I knew the next second the air would come hissing out of the Zodiac." Fortunately, the patch he had pulled off wasn't crucial. But clearly both

Lippoth and Adams had gone over the edge. "He had no idea what he had done. His eyes were flat and dull." Soon Lippoth and Adams were having a conversation that left Cavanagh and Kiley mystified. Lippoth asked Adams if he wanted a smoke, and Lippoth said, "Sure . . . I got some sandwiches. You want one?"

According to Nathaniel Philbrick's *In the Heart of the Sea*, a history of the whaling ship *Essex*, on which Melville's *Moby-Dick* was based, "It is not uncommon for castaways who have been many days at sea and suffered both physically and emotionally to lapse into what has been called 'a sort of collective confabulation,' in which the survivors exist in a shared fantasy world. Delusions may include comforting scenes from home."

Lippoth's brain, overloaded with sodium, told him that he was in an entirely different environment. And the result of losing that connection with his world killed him. The feeble, flickering circuitry of his brain seized on whatever was close at hand in recent memory. He was back on *Trashman* now, and they were at the docks. It was so vivid it literally blotted out the evidence of his senses. Lippoth acted on that imagined world. He began talking about going to get the car so they could unload. "You guys bring the boat in and I'll get the car." He climbed over the side of the Zodiac into the water. "I'll be back in a few," he said matter-of-factly. The sun was setting as he hung there, clinging to the boat. "I can't take this anymore," he said. "I'm going to go get the car."

Then he swam away. Cavanagh and Kiley stared in horrified disbelief. They briefly discussed whether they should try to go get him, but it was already too dark, and they knew the sharks were everywhere. As they watched him ride over the swells, they heard "a terrible, gut-twisting scream," as Lippoth was eaten alive.

Remarque reported a remarkably similar incident in *All Quiet on the Western Front* of a young soldier reacting to combat:

I have been watching him for a long time, grinding his teeth and opening and shutting his fists. These hunted, protruding eyes, we know them too well. . . . He had collapsed like a rotten tree.

Now he stands up, stealthily creeps across the floor, hesitates a moment and then glides toward the door. I intercept him and say: "Where are you going?"

"I'll be back in a minute," says he, and tries to push past me.

Which would not have been so extraordinary except for the fact that they were in a bunker, and outside that door "the bombardment extends undiminished as far as the artillery lines."

In the middle of Kiley's night of horror, Adams got on top of Meg, who was near death, covered with sores and horribly infected wounds. "Hey, lady," he said, "I think it's about time you and me had sex." When she started weeping, Adams screamed: "Fuck you. Fuck you. I'm tired of playing games. I'm going back to the Seven-Eleven to get some cigarettes." Then he, too, slipped over the side. Soon, they could actually feel the sharks eating Adams beneath the dinghy. "The bow lifted up, then dropped back down," Kiley reported. "The dinghy spun around. I was paralyzed with terror."

Years later, Kiley told me: "That is the only time in my life I felt like I was just walking down that fine line of sanity. Here's Mark, he's overboard, the sharks are eating him, Meg's dying, it's in the middle of the night, and it's the first night that the stars are out. It was just very haunting." Noticing the stars, the beauty of the sky, was a significant sign of strength. For although Kiley walked the line, she did not slip over it as Lippoth and Adams had done. Instead, she continued to struggle to consolidate her inner strength, to adapt and survive. A crucial moment for all survivors comes when they become convinced that they will survive. Often it occurs after a spiritual experience of the beauty of the world. "When I was out there and people started to die," Kiley told me, "I knew that I wasn't going to die. It was just a matter of figuring out how."

She concentrated on the sky and on her praying. "Focus on the sky," she told herself, "on the beauty there."

Once again she experienced the hardness survivors need in order to carry on amid the horror. "I knew Meg was going to die soon. And God forgive me, I felt relief. She was the last weak link. Once Meg was gone it would be Brad and me. We would be at square one, able to concentrate completely upon staying alive. And at that moment I knew with startling and absolute certainty that I was not going to die." Facing down horror, finding beauty, then feeling certainty: every survivor describes the same pattern. In the stages of dying, the last step is acceptance. In survival, it is total commitment.

Every survivor I spoke with said that there was a moment of revelation in which he was suddenly seized by the certainty that he would live. Mary Jos was on the seventy-eighth floor of the south tower of the World Trade Center when the wingtip of United Flight 175 sliced through the sky lobby, killing nearly everyone standing there waiting for the elevators with her. She was unconscious for a time, burned and bleeding. When she came to and saw the destruction and fire around her, she thought of her husband and told herself, "I am not going to die." She got out.

IT HAD taken Kiley three days, but she had made the successful transition from victim to survivor. She and Cavanagh took turns sleeping. She felt the calm and control of the rescuer. "I would stay here like this, watching him until he woke up, making sure he didn't drown in the swill." Now they were on the voyage of survival. Paradoxically, they didn't need to be rescued. Certainly, their bodies would die sometime, perhaps soon if they weren't rescued. But they'd come to terms with their world.

There is an old Zen story about a young man who passionately wants to become a master swordsman. He goes to the Kundo master and begs to be taught, but the master puts him to work in the garden instead. Every time the student isn't looking, the master sneaks up behind him and whacks him with a stick. The student is terribly frustrated, and this goes on for months, then years. No matter what the student does, he can't seem to sense when the master is behind him, and he is constantly covered with bruises. But then one day, the student is in the garden, as usual, hoeing and weeding, and the master sneaks up behind him. He swings. The student ducks. The master misses. The student is overjoyed. He leaps up and shouts, "Now will you teach me swordsmanship?"

"Now you don't need swordsmanship," the master says.

When I first heard that story, I almost wept, because it seemed so much like me and my father.

Meg Mooney died at last, and they took her shirt for extra clothing and stripped off her jewelry to return it to her family. Then they let her slide over the side. That beautiful young woman, who had been so stylish and bubbly in her sunglasses and bikini just days before, floated facedown until she drifted out of sight, returning her borrowed materials to the silver sea.

By the time they saw another ship, Kiley had completely accepted her new world. Cavanagh wanted to try to signal, but she told him, "Don't waste your energy. It'll never see us."

Maurice and Maralyn Bailey had a remarkably similar response after three and a half months adrift. They saw a ship and Maralyn wanted to signal. Maurice said, "Let it go on . . . this is our world now on the sea, amongst the birds and the turtles and the fish." As Dostoevsky wrote in *Memoirs from the House of the Dead*, "Man is a creature who can get used to anything, and I believe that is the very best way of defining him."

Survival is a simple test. There's only one right answer, but cheating is allowed. And in the end, Debbie Kiley passed. A Russian freighter picked her and Cavanagh up after five days. It's easy to die without water in three. Their Positive Mental Attitude kept them both alive.

I asked her what advice she had for others. "Trust your gut," she said. "I had my misgivings about the trip all along. It just didn't feel right. So I have just one piece of advice for people: Your gut tells you what to do. Believe it. I didn't, and a lot of people are dead, and I have to live with it. Also, never forget that you can't depend on anybody. You really have to have it within yourself to do it."

TWELVE

A VIEW OF
HEAVEN

ON HIS FIRST full day adrift, Steve Callahan began a diary: notes on his health, the condition of the raft, and the status of his supplies. He was orderly. He set small tasks. He took responsibility and focused tightly. But he never lost his sense of humor. "I continue to make light of whatever I can in order to relieve the tension." Leadership, order, and routine are all important elements of survival. Callahan had established all three before twenty-four hours had passed.

Even so, at that early stage, his hold on himself was fragile. From time to time, he'd succumb to attacks of emotion, which he had to fight to control. Yet his grip on reason, his ability to think clearly and in a practical way, is remarkable. "Desperation shakes me. I want to cry but scold myself. Hold it back. Choke it down. You cannot afford the luxury of water wept away. . . . Survival, concentrate on survival," he told himself.

Callahan would think of home and future plans and happy times, especially meals. But it was early in his ordeal, and he didn't yet know that such thoughts were normal. He'd scold himself to stop. He was a hard taskmaster.

John Leach uses the term "active-passiveness," meaning "the ability to accept the situation one is in but without giving in to it. . . ." In a survival situation, there is a lot of waiting, and there should be a lot of resting as well. To do that well takes self-control and training, but learning to wait can also help to prevent boredom and hopelessness from setting in. Jackie Mann, one of the Westerners kidnapped by Arab terrorists in Beirut, found that a big part of his job was to learn his captors' capacity for waiting and patience.

Ultimately, it is the struggle that keeps one alive. What seems a paradox is simply the act of living: Never stop struggling. Life itself is a paradox, gathering order out of the chaos of matter and energy. When the struggle ceases, we die. Scientists have long observed the seeming mystery: You can will yourself to die. So it was just as important that Callahan desire to be somewhere better as it was that he snap himself back to the reality of his situation. Gradually, he understood: "But the desire to dream lingers. It is my one relief." True, you can't give yourself up to it. But you can use it and even enjoy it. Survival depends on utility, but it also depends on joy, for joy is the organism telling itself that it is all right.

Paradoxically, his personality was disintegrating and consolidating at the same time. He was noticing the split all survivors feel, and he notes in his book that "the distinction between the [mental and physical] parts of myself continues to grow sharper." He finds that he must coldly force himself to think rationally and force himself to work. "I try to comply with contradictory demands, but I know the other parts of me have bent to my cold, hard rationalism as best they can. I am slowly losing the ability to command, and if it goes, I am lost. I carefully watch for signs of mutiny within myself."

Callahan recounts a conversation he had with his "captain." He asked for water, and the captain said, "Shut up! We don't know how close we are. Might have to last to the Bahamas. Now, get back to work."

"But, Captain—"

"You heard me. You've got to stay on ration."

Nearly all survivors report hearing what they call "the voice." It tells them what to do. It is the speaking, rational side of the brain, the one that processes language, the wellspring of reason.

Callahan had a huge advantage over most recreational survivors because he had already had such experiences in the course of following his passion for sailing: He had learned, perhaps unconsciously at first, to effect the split while sailing, dividing himself into what Joseph LeDoux calls "The Mental Trilogy" of emotion, cognition, and motivaton. "When I am in danger or injured, my emotional self feels fear and my physical self feels pain. I instinctively rely on my rational self to take command over the fear and pain." But, under extreme stress, he begins to find that what he had experienced as three selves working together are becoming dissociated. "My rational commander relies on hope, dreams, and cynical jokes to relieve the tension in the rest of me."

The sort of discipline he describes is part of the process of wrenching some sort of order out of chaos. It is the basic process of life, only here with much more energy at stake because the system is being driven so much harder than it normally is. He set his water ration at half a pint each day. "To take only a mouthful every six hours or so is difficult discipline. I have decided not to drink seawater."

Pondering the whale that may have sunk him, Callahan recalled other whales he'd seen at sea. "Suddenly there's a huge beast there. And at that moment a profound emotion—not fear—rises from the depths of my soul. It is like seeing a friend whom I thought I'd never see again, who miraculously appears. . . . I feel a wonderful electricity in the air, an aura of immense intelligence and sensitivity." I recalled the lifeguard Mike Crowder in Kaua'i telling me about the *mana* coming off a surfer as he walked past. You hear things like that, and at first they sound silly. But you hear enough survivors of impossible situations say it, and you become a believer. Cynicism isn't in the lexicon of nature. Animals, natural forces, are frank. Survivors therefore must have spirituality and humility. As Callahan puts it: "I believe in the miraculous and spiritual way of things. . . . I do not know the true workings of that way. I can only guess and hope that it includes me."

All the while, Callahan's cognition was firmly in control of his actions. Hunger is one of the most powerful of all emotions. The survival experts call it "food stress." Although the body can go perhaps three weeks without food, the emotional drive is immense and becomes an obsession, a drive as strong as any fight or flight response. It can't be ignored.

People, of course, will kill for food. Remarque wrote: "The corned beef over there is famous along the whole front. Occasionally it has been the chief reason for a flying raid on our part." My father told me of lying broken and near death in the Nazi prison camp and still having elaborate food fantasies.

Major Gene M. Lamm, an Army physician who was a Korean prisoner of war, put it this way in John Leach's *Survival Psychology*: "Regardless of what it is, eat it. One of the basic principles of survival medicine is to eat. . . . If you miss one meal as a prisoner, it will take you weeks to regain your lost strength." He described maggots found in rotten fish as being "quite good."

Callahan had salvaged some raisins from *Solo*, but they became waterlogged and rotted. "The stems are slimy and bitter. I eat them anyway."

When Callahan realized that he might use his spear gun (by chance, he'd bought it in the Canaries and stashed it in his survival pack) to catch the dorados that swarmed around his raft, he went a little crazy with craving. But he stopped himself before taking action. "Wait . . . what if it is a strong fish? I hurriedly tie a piece of line through the gun handle and onto the raft. My stomach growls. Four days on a pound of food. I'm trembling with excitement." He remembered to monitor the sea so that his weight at the edge of the raft didn't capsize him. Then he took his time. The spear gun was not much more than a toy, but it saved his life.

Callahan began talking to the fish. "You get into some really crazy thinking," Ken Killip told me thirty months after his own rescue from Rocky Mountain National Park. But it's not crazy, it's normal survivor behavior. Killip began feeding his rations to the marmots during the five days he was lost just to get them to come close. He talked to them as well.

Christopher Burney kept a pet snail in Buchenwald. "When the soup came, I gave the snail little pieces of cabbage." So the behavior Tom Hanks portrayed in the movie *Castaway*, talking to a volleyball, while it was taken a bit too far by the Hollywood types, was still based on reality: the acceptance of our world. We are social animals, and whoever is present is our society. We must love, and what we love we take into our bodies. Callahan loved his dorados even as he ate them.

Survivors develop a mantra. It can be anything. Callahan's seems to have been nothing more than the word "survival" itself. Over and over again, he'd say something like, "Concentrate on *now*, on survival."

Yossi Ghinsberg, the Israeli hiker, wrote that, in the early moments of impact when his raft was swept away in a raging river, spitting him out and stranding him for three weeks in the Bolivian jungle, he began telling himself, "*Don't cry. Don't break now. Be a man of action.* . . . When I found myself feeling hopeless, I whispered my mantra, 'Man of action, man of action.' I don't know where I had gotten the phrase . . . I repeated it over and over: A man of action does whatever he must, isn't afraid, and doesn't worry."

Callahan, like the rest of us, was dogged by faulty technology: his solar stills (devices for distilling water from the moisture in the air) wouldn't work. The inner-tube raft began cracking long before he was through with it, and the waterproof covering had begun to disintegrate. The rain poured in on him, and what water the canopy could catch was contaminated and not drinkable. "Trying to swallow it is like forcing down another man's vomit. . . . I curse it." Who makes those things? Certainly not someone who has tested them in the ocean for months at a time. And his was a six-man raft, considered extravagant for his solo voyage. He couldn't even lie down in it but had to sleep curled up. Only by constant and very clever tinkering was Callahan able to survive using his so-called professional survival gear. *Caveat emptor.* But survival is an act of art and craftsmanship. It involves the order of craft and the spontaneous invention of art.

The more Callahan accepted his new world as normal, the more he embraced it with the willing, thinking part of his mind. The trick is to own the world and to let it own you. To come, quite literally, to terms. Hence, he named the dorados his "doggies." He called the triggerfish "butlers" because "they have that starched collar look." He named his raft *Rubber Ducky*. And the place where he hung up the strips of fish to dry he called, "the butcher shop."

But part of accepting the world is retaining selfhood, knowing your own inner needs, especially the need for rest and relief. Even in short-term survival situations, memory becomes a tremendous resource for salving the spirit, finding place.

Remarque describes how, in the midst of a mortar attack,

I see a picture, a summer evening, I am in the cathedral cloister and look at the tall rose trees that bloom in the middle of the little cloister garden where the monks lie buried. . . . A great quietness rules in this blossoming quadrangle, the sun lies warm on the heavy gray stones, I place my hand upon them and feel the warmth. . . . The image is alarmingly near; it touches me before it dissolves in the light of the next star shell.

Callahan wrote: "The sun sizzling overhead roasts my dry skin. . . . I envision myself sprawled on an Antiguan beach. In a moment I will rise and fetch a cold rum punch—no need to yet, plenty of time." He recognized the need for escape. "I want desperately to keep these other worlds safe from pain and depression so that I can escape to them whenever I wish. My own propaganda is intoxicating."

A survival situation brings out the true, underlying personality. Our survival kit is inside us. But unless it's there before the accident, it is not going to appear magically at the moment it's needed. When you consolidate your personality as a survivor, what you get is the essence of what you always had. A survival situation simply concentrates who you are. It drives the natural system you've developed over a lifetime, and it drives it harder. Whether or not it becomes chaotic at the boundaries depends on what you've put into it over a lifetime. Your experiences, education, family, and way of viewing the world all shape what you would be as a survivor.

After experiencing a hurricane as a child, Callahan put together a survival kit with money, a knife, fishing tackle, and a few other things. "If anyone survived, it would be me," he told himself. Also, he was once bedridden due to blood poisoning when he was a teenager but exhibited a survivor's attitude even at the age of sixteen. "Instead of dwelling on my ailment, I told myself that at least I still had a clear head, strong arms, and one good leg."

Survival had always been part of Callahan's way of life. Those are the people who survive best. Others require rescue and a great deal of luck. In the hurricane, in the sailing he'd done, Callahan had experienced forces that could kill him, so he was a believer. He had an artistic, craftsmanlike temperament. He had patience and humility. He was meticulous, a perfectionist. He had spent a lifetime building a survival raft that didn't leak.

Survivors find wealth and happiness in the smallest things. They never say, "Why me?" or, "I can't take it anymore." When his raft leaked and he was sinking, with seawater destroying his precious cache of dried fish, Callahan spent eight days tinkering together a patch that worked. He drove himself to complete exhaustion and had to beat off shark attacks at the same time. But coming up out of sleep the morning after his thousandth failure, a voice in his head told him: "I want to LIVE, to LIVE, to LIVE!" When at last he succeeded, after many more hours of maddening work, as well as divine mechanical and engineering inspiration involving a fork, no less, he wrote, "My body hungers, thirsts, and is in constant pain. But I feel great! I have finally succeeded!" He wasted no energy on anything but positive, forward motion. He rarely slipped back or questioned the fates. (Only occasionally did he consider giving up what he called his "pointless struggle.") Most of the time, he rose to each challenge with a big heart bursting with enthusiasm. It was his fifty-third day adrift.

Like all true survivors, Callahan was doing it not just for himself. Survivors are always connected to loved ones, friends, society. They survive because they are rescuing the species, not just themselves. It's another paradox of survival: The individual doesn't matter. But the survival instinct of the individual must matter if the species is to survive. That's one reason survivors do it for another.

Callahan was sustained by thoughts of his mother: "You had better not give up that easily," his mother told him once when she was worried about his sailing habit. "You have to promise me to hang on as long as you can."

I remember asking my father what his first thoughts were when his plane was hit. I was probably ten years old. He told me he thought of his mother, then my mother, and the pain they were both going to feel. He said he was intent on the question of how he was going to get back to them to prevent that pain.

Callahan constantly reminded himself of the basic things, the important things: "The important thing is to keep calm. . . . I can only afford success. Don't hurry. Make it right. . . . Patience is going to be the secret, and strength."

At the heart of his magnificent survival is what I had gone looking for, the beautiful thing contained in that dull military locution, Positive Mental Attitude: "I am constantly surrounded by a display of natural wonders. . . . It is beauty surrounded by ugly fear. I write in my log that it is a view of heaven from a seat in hell."

The Stoics are the best survival instructors. As Epictetus wrote, "On occasion of every accident (event) that befalls you, remember to turn to yourself and inquire what power you have for turning it to use." The secret is to meet adversity head-on. When his spear gun lost its elastic motor, which spelled death for Callahan if he couldn't fix it, he reflected, "It's always a challenge to try and repair an essential system with what one has at hand. . . . In many ways, having a jury rig succeed is often more gratifying."

Callahan provides a rare example of the completely consolidated personality:

In these moments of peace, deprivation seems a strange sort of gift. . . . My plight has given me a strange kind of wealth, the most important kind. I value each moment that is not spent in pain, desperation, hunger, thirst, or loneliness. Even here, there is richness all around me. As I look out of the raft, I see God's face in the smooth waves, His grace in the dorado's swim, feel His breath against my cheek as it sweeps down from the sky.

He sees a rainbow and writes, "I feel as if I am passing down the corridor of a heavenly vault of irreproducible grandeur and color."

He saw that to lose everything at the edge of such a glorious eternity is far sweeter than to win by plodding through a cautious, painless, and featureless life. And that, of course, is why people undertake adventures such as solo voyages of the Atlantic to begin with. The true survivor isn't someone with nothing to lose. He has something precious to lose. But at the same time, he's willing to bet it all on himself. And it makes what he has that much richer. Days stolen are always sweeter than days given.

BY THE sixty-fifth day, Callahan was advancing closer and closer to death. His amazing abilities, strength, and determination had slowed the process considerably, but no one lives forever. Not even the best survivor. Sometimes the process can go on for an ungodly long time, as was the case with Shoichi Yokoi, a sergeant in the Japanese Imperial Army, who hid out in the jungle when the Americans won the island of Guam in 1944. Telling himself, "I am living for the Emperor," he survived on his own until he was found in 1972.

When at last Callahan reached the Caribbean Islands, on April 21, 1982, having made a successful solo crossing of the Atlantic (minus his boat), a fishing skiff was attracted to his raft because of the birds flying over it, who were after the school of dorados Callahan had herded across the Atlantic. The fishermen, seeing his condition, offered to take him to land. He said, "No, I'm O.K. I have plenty of water. I can wait. You fish. Fish!" And he waited, drinking his hoarded water (by that time, he had plenty) and watching their joy in fishing those great shimmering dorados. He drank five pints while watching, having stayed on a ration of half a pint a day for seventy-six days.

THIRTEEN

THE SACRED
CHAMBER

TWO BRITISH MOUNTAINEERS, Joe Simpson and Simon Yates, had decided to make a first ascent of the 4,500-foot southwest face of the 21,000-foot Siula Grande, a mountain in the Cordillera Huayhuash of the Peruvian Andes. They achieved the summit in May 1985 and were on the very difficult descent when Simpson broke his leg. He was above 19,000 feet on a completely uninhabited, snow-covered mountain in Peru. He knew that Yates couldn't pull him up on the rope that connected them; it was an impossibility of physics. Even as Simpson hung there, upside down after his fall, he thought, "I've broken my leg, that's it. I'm dead. Everyone said it . . . if there's just two of you a broken ankle could turn into a death sentence."

At first, as he described later in his book *Touching the Void*, Simpson began to follow the common pattern of reaction to catastrophe: denial. He told himself that the leg wasn't really broken, it was only sprained. That thought came unbidden, even as he was seeing "the kink in the joint and knew . . . the impact had driven my lower leg up through the knee joint."

Simpson was splitting into two people: one perceived the reality, the other denied it—reason and emotion. Then hot waves of emotion poured over him, and tears of self-pity streamed down. Within moments, he had gone through denial to anger: "I felt like screaming, and I felt like swearing. . . ." You mustn't linger too long in any but the last of Kübler-Ross's stages, which is a yin-yang stage that can be either resignation or determination.

Joe Simpson had to do something to stop his descent toward death. He knew he had screwed up, screwed up big time. Then something in him clicked. He "stayed silent. If I said a word I would panic. I could feel myself on the edge of it." He stopped the progression through the stages of dying at that moment. He took cognitive control. He had successfully put his thinking brain in balance with his emotional brain so that the two could help him to function. He didn't know how he was doing that. He didn't have to know. Neither did he know what Yates would do; probably leave him. Simpson simply took charge of the things within his power, namely himself, and left the rest behind.

By the time Yates peered over the ridge, Simpson was right side up (both mentally and physically) and calmly told his partner, "I've broken my leg."

Although Simpson could see the change in Yates's expression, all Yates said was, "Are you sure it's broken?"

"Yes."

Without batting an eye, Yates said, "I'll abseil down to you," and began working on an improvised anchor for the rappel. Of course, dozens of thoughts and emotions were powering through both their minds at that point, and Yates knew, too, how close to death Simpson was. His immediate thought was, "You're fucked, matey. You're dead."

But neither of them said a thing. They simply set to work. Yates wasn't going to leave him. He, too, was going to do what was in his power, nothing less, nothing more. "It was all totally rational,"

Yates said later. "I knew where we were, I took in everything around me instantly, and knew he was dead. . . . I accepted without question that I could get off the mountain alone."

Al Siebert, in his book *The Survivor Personality*, writes that "The best survivors spend almost no time, especially in emergencies, getting upset about what has been lost, or feeling distressed about things going badly. . . . For this reason they don't usually take themselves too seriously and are therefore hard to threaten."

Yates rappeled down to Simpson and examined him. He gave him some pain medicine. Neither spoke. As Simpson put it, "In an instant an uncrossable gap had come between us." They were both working on shifting decisions out of the amygdala and into the neocortex. In such a dire emergency, the amygdala would urge instant action without thought. It has the chemical authority to do that, too. So it takes energy, balance, and concentration to shift control to the executive functions of the neocortex. Simpson and Yates didn't want to upset that delicate balance with any extraneous inputs.

Simpson understood it, too, and explained, "It felt as if I was holding something terrifyingly fragile and precious." They both instinctively recognized their need for privacy in their individual struggles for survival, those critical moments of consolidation. Each man must do his own surviving. He may do it by rescuing someone else, by not being a victim, but just as only he can live his life, so too his survival cannot be shared. No one can do it for him. Privacy is important for strength in that lonely work.

Remarque wrote about going home on leave. "[F]ather wants me to tell him about the front . . . he does not know that a man cannot talk of such things; I would do it willingly, but it is too dangerous for me to put these things into words." And of the "uncrossable gap," Remarque comments, "There is a distance, a veil between us." He knew that he was not really at home. He had to go back to the front and survive.

. . .

WHEN YATES had finished checking Simpson, he discovered that
his own rope above had caught on something, so he had to climb
back up to retrieve it. He recovered and adapted quickly. He saw
opportunity in adversity. "In a way, it took my mind off things,
and gave me time to settle into the new situation."

Time was of the essence. They had run out of water and fuel.
They had to get down fast or they would both die. So the process
of coming to terms with the new world had to take place promptly.
They had clamped down hard and sure on the basic tools of survival. Their attitude was cool, open, empathic, and committed.
Their approach was orderly, bold yet cautious, inventive; in a
moment, it would become playful as well.

As Yates began his dangerous, unroped climb to the top, as he
later put it, "I left him then, and forgot about him." Cool turned to
cold. That's what it takes. He was fully committed.

It is difficult to convey the objective hazard of Simpson and
Yates's situation. They were on a snow cliff that wasn't solid
enough for the use of conventional ice-climbing tools, such as
axes. Coming off the summit, they'd had to plunge-step into it,
burying their feet and burying their ice axes, including their arms,
just to keep from falling. Even then, the snow gave way, crumbling
with their every movement. As they placed their feet on the
descent, at each step the snow would compress, and they had no
idea when one such slide would simply continue, accelerating into
a fall that would have taken them several thousand feet straight
down. Now the slightest mistake could result in death for Simpson,
if not Yates as well. Yet they would be called upon to do the most
outrageously risky things and to do them with the utmost caution.

Simpson had unroped so that Yates could set up the rappel. Now,
with Yates once more beside him, clinging to the face, as Yates
tells it, "Joe tried moving beside me and very nearly fell off. I
grabbed him and put him back into balance. He stayed silent . . .

he knew that if I hadn't grabbed him he would have fallen the length of the East Face." Indeed, "The climb up the edge of the cliff was the hardest and most dangerous thing I'd ever done." Yates arrived at the top "shaking and so strung out that I had to stop still and calm myself." But it was that very ability to remain calm that made what they were about to do possible.

As Yates watched, Simpson began traversing upright on the nearly vertical wall of snow. Yates later said, "He moved so slowly, planting his axes in deeply until his arms were buried and then making a frightening little hop sideways. He shuffled across the slope, head down, completely enclosed in his own private struggle. . . . It occurred to me that in all likelihood he would fall to his death. I wasn't disturbed by the thought. In a way I hoped he would fall. I knew I couldn't leave him. . . . If I tried to get him down I might die with him."

Simpson was learning what it means to be playful in such circumstances: "A pattern of movements developed after my initial wobbly hops and I meticulously repeated the pattern. Each pattern made up one step across the slope and I began to feel detached from everything around me. I thought of nothing but the patterns." His struggle had become a dance, and the dance freed him from the terror of what he had to do.

ON APRIL 26, 1976, Lauren Elder was a twenty-nine-year-old painter living in Oakland, California. Her boyfriend's boss, Jay Fuller, had a four-seat Cessna and had decided to take Elder and his own girlfriend over the mountains to Death Valley for a picnic. Owing to a mistake in navigation, Jay crashed his plane at 12,360 feet on a peak half a mile south of Mount Bradley in the Whitney Quadrangle of the Sierra Nevada. His girlfriend died quickly, and Jay died the next day. Elder was stranded on the icy cliff.

As she recounted in her book *And I Alone Survived*, in the immediate aftermath of the crash, even as Jay and his girlfriend were dying, "something peculiar happened. I got angry. It came simmering up from inside. I thought: That's not going to happen to me. I am going to get down from here. That's not going to happen to me. Anger displaced inertia."

She displayed all of the typical features of a successful survivor. Using gasoline from the plane, she kept a fire going all through the icy night. She drank beer that Jay had brought along, to keep from being dehydrated. And starting off the next day, wearing high-heeled boots, a short skirt, and no underwear, she climbed and hiked for thirty-six hours to reach a small California town.

She told me, "I kept stopping to appreciate how beautiful the place I was in was. There I was in this amazing wilderness, and I had the whole place to myself." She even found time to stop and skinny-dip in an icy mountain pool and was moved to cry out with joy. Survival is the celebration of choosing life over death. We know we're going to die. We all die. But survival is saying: perhaps not today. In that sense, survivors don't defeat death, they come to terms with it.

Elder didn't require rescue, didn't further injure herself (she'd broken her arm in the crash), and did get out alive. When I spoke to her, she talked of using patterns, as Joe Simpson had. She turned her grueling hike into a dance of joy. Countless survivors have reported the same thing: by developing a pattern and then fixing on nothing but making the pattern perfect, they were able to get out of seemingly impossible situations.

Chanting, for example, is a pattern that can alter consciousness and calm the mind. The military uses marching and songs to move troops and keep fatigue and emotions under control. The brain is organized to recognize patterns, language and numbers among them. They are elemental. Life is, after all, the order inside the chaos. Everything else goes down with entropy. We rise up. And we do it with patterns. Patterns of proteins. Rhythms of breathing and heartbeat, patterns of day-to-night, lunar and menstrual cycles, the regular marks on bone fragments that the first people left behind. To make a pattern, to use rhythms, means quite literally to live.

Extraordinary effort in human work has always taken rhythmic play along as its helpmate. It would not be surprising to learn that the first people to cross the Bering Strait danced across, singing a song, even as slaves who were forced to build railroads sang chaingang songs to the ringing of a 9-pound steel. So Joe Simpson had come to accept his situation; he employed an inventive concept to reach the stage of playfulness; and he used that to create a rhythmic pattern to achieve a short-term goal.

"I knew I was done for," Simpson wrote later, "It would make no difference in the long run." But he kept going anyway because of the pattern.

Yates appeared on the wall near Simpson, then moved ahead of him to cut a trench that would make Simpson's traverse easier. Simpson continued to repeat his pattern along the trench. He thought, "If I asked Simon to help, I might lose this precious thing. . . . I remained silent, but it was no longer for fear of losing control. I felt coldly rational."

Epictetus said, "And let silence be the general rule, or let only what is necessary be said, and in few words. And rarely and when the occasion calls shall we say something." Tom Wolfe wrote in *The Right Stuff*, "One of the greatest sins was 'chattering' or 'jabbering' on the radio. . . . A Navy pilot (in legend, at any rate) began shouting, 'I've got a MIG at zero! A MIG at zero!'. . . . An irritated voice cut in and said, 'Shut up and die like an aviator.'" Significantly, an aircraft carrier operation in real combat is called a Zip Lip.

Simpson was so focused on his patterns that, when Yates asked how he was doing, Simpson had forgotten his partner was even there. He lost track of time, too. "I had almost forgotten why I was doing [the patterns]." Only when he was off the cliff to one side did he allow himself the luxury of being afraid once more.

Since the accident, they had each done everything right. They had discovered inside of themselves everything a great survivor needs. But they both knew theirs was a precarious position and a fragile state of mind. They were "holding something terrifyingly fragile and precious."

Now they found themselves on a very steep slope of unconsolidated snow. They worked out a system whereby they would dig a bucket seat in the snow for Yates, and he'd lower Simpson on two 150-foot ropes tied together. Then he'd climb down to Simpson, dig another seat, and start over. To succeed, they'd have to repeat that pattern perfectly ten or twelve times.

Recognizing the desperation of their circumstances, Yates lowered Simpson very rapidly. As he did so, Simpson's broken leg caught again and again on the snow, causing him excruciating pain. But neither ever said, "I can't take it anymore." Neither said, "Why me, Lord?" There was no self-pity.

Instead of complaining, each time Yates came down to where Simpson waited, digging him another seat, Yates grinned at him. Simpson reported that, "his confidence was infectious." As Simpson put it, "We had lost the sense of hopelessness that had invaded us at the ice cliff." And all the while, the all-important struggle was taking place in Simpson's mind between two voices. The more battering his leg took, the more one voice begged him to "Rest, leave me be!" But, "we had locked ourselves into a grim struggle." He stayed inside himself, never looking outside for someone to blame. Instead, his rational, cold voice said, "I could adjust to the steady pain."

Weather conditions were worsening; time was running out, and they were both experiencing hypothermia and frostbite. Their hands were becoming useless. Simpson came to the end of the rope once again and began digging the seat for Yates. To his horror, he discovered solid water ice beneath a small amount of snow. "It was a measure of how cold I had become to see how long it took before I thought of hammering an ice screw into the slope," he wrote. Once hypothermia sets in, despite the best efforts, cognition deteriorates quickly. To use Jack London's phrase, "The wires were down." Even when the idea at last occurred to Simpson, it took a monumental effort to turn the correct idea into action. He managed to hook the wires back up again.

After pounding in the screw as an anchor for the rope, he started exercising to warm himself. That level of cognitive functioning under such circumstances is truly remarkable. At each new pitch, Yates knew how much he was hurting Simpson. He said later, "It was strange being so cold about it"; but it was efficient. That was all there was now, to be as efficient at surviving as nature itself was being at taking their lives.

Seeing the water ice into which Simpson had pounded the screw, Yates understood that water didn't simply appear in one spot. It had flowed there. The ice would have to be part of a larger formation. There must be something beneath them on the slope. They couldn't see because of the weather. They decided to rappel.

After successfully completing the rappel, Yates and Simpson looked at each other and Simpson later said, "It felt like some cliché from a third-rate war movie." Indeed, it was. That's why Hollywood movies are structured the way they are: It's the reason that heroes in movies are cool and show no emotion while the villains go berserk. And the reason movies have three acts is because survival comes in three acts. (In Act One, you get into trouble; in Act Two, you struggle; in Act Three, you get out.) Simpson and Yates were heading for the "Act Two curtain," as it's called among Hollywood scriptwriters, that point at the end of the second act where the hero is farthest from his goal and it appears as if he's done for. In Act Three, with a superhuman struggle and a big chase scene, he overcomes impossible odds and finds redemption and triumph. I've done some Hollywood screenwriting, though not very successfully, and I always wondered why we had to write our scripts to that formula. I obviously knew nothing about why there are movies in the first place.

They were so focused now, Simpson didn't notice that night had fallen. (Or, as Mike Yankovich put it, "You're at a quarter mile and someone asks you who your mother is, *you don't know*. That's how focused you are.") Everything was working the way it was supposed to, when one of their mistakes, long forgotten, caught up with them. Zooming down the slope, ignoring the pain, Simpson went straight off the cliff. Having not exactly planned the descent, they had somehow missed that 800-pound gorilla. Simpson now dangled hopelessly over the cliff, unable to reach the ice wall 6 feet away to plant his ax.

Far above, Yates's senses were so finely tuned that he could read every trembling vibration in the rope, and now he could tell what Simpson was doing. Or rather what he was not doing: He wasn't climbing or standing on anything. He was hanging.

This is precisely where they had started. There was no way Yates could pull Simpson up, but now there was also no way for him to get to Simpson. Simpson could neither climb nor descend. Neither could see the other or communicate. Simpson knew Yates's seat was crumbling. None had lasted much longer than the time it took for each of Simpson's rapid descents. Checkmate. The greatest chess master, Mother Nature, does not resign. There was nothing Yates could do, and as his seat let go and he began to fall, he cut the rope. Even then, he did not succumb to emotion. "I was actually pleased that I had been strong enough to cut the rope. . . . A lot of people would have died before getting it together to do that!" And after all, this isn't about heroism, it's about survival. Once again, Yates had done what was necessary. If it was murder, it was premeditated. He'd even thought to untangle the rope before cutting the line so that it wouldn't catch his foot and pull him down. He was sure Simpson was dead and that he had killed him, but "I felt cold and hard."

SIMPSON FELL and fell for a very long time. The world went black. He landed with a huge concussion. When he came to his senses, he was in a deep crevasse. He had punched through a snowy roof and landed on a snow bridge. His pack and the snow had broken his fall. Aside from his leg, he was miraculously uninjured. But the miracle had left him more stranded than before. Thinking that Simpson was dead, Yates would leave the mountain. It was the Act Two curtain.

And yet, when Simpson realized that he was alive, he laughed. A key first move. That the fall hadn't killed him was miraculous; that it hadn't killed his spirit was more so.

He lay in a precarious position on a slick and snowy bridge above a long drop into darkness. Simpson began to analyze his situation. Careful not to slide off, he got new batteries out of his pack and put on his headlamp. Working by that light, he managed to squirm around and put an ice screw into the wall nearby. He could see no floor, only blackness below. High above, however, he could see the hole he had punched and a sky furious with stars. As he fought off periodic bursts of emotion (laughter, weeping), Simpson's "calm rational voice" spoke to him. The split was working again. "The rest of me went quietly mad while this calm voice told me what was happening and left me feeling as if I were split in two." As Simpson had done when he first broke his leg, he spent some time swearing and cursing out loud, his voice echoing off the crystal walls of his ice tomb. Instead of giving in to fear, he used anger to power his actions. But the hole far above him "was as unreachable as the stars."

At last, Simpson calmly accepted his new world, even while knowing it was impossible to climb out. (Epictetus again: "Lameness is an impediment to the leg, but not to the will. And add this reflection on the occasion of everything that happens; for you will find it an impediment to something else, but not to yourself.") So he tried to climb.

He was right. It was impossible.

Simpson settled on his only other option: lowering himself into the black and unknown abyss. If it was farther down than his rope

length (whatever that was now that it had been cut), he'd go off
the end and die. He intentionally did not tie the safety knot,
because he didn't want to hang there for several days or even
hours, dying slowly, and he knew he'd be unable to ascend.

Simpson was experiencing what Leach calls the paradox of
"resignation without giving up. It is survival by surrender."

Despite frozen hands, he managed to tie into his ice screw. He
hooked in and slipped off his perch with terror pouring through
him, barely under control. He was fully committed now to survive
or die. Several times on his seemingly eternal rappel, Simpson was
seized by panic and simply stopped, hung there, and found himself
unable to go on. His mind was swamped by the image of feeling
and hearing the rope whiz through the rappel device as he
plunged into the bottomless pit. But he managed to keep going,
and before the rope ran out, he found himself staring at a floor in
the pale yellow light of his lamp. He just hung there for a
moment, taking it in. He was aware that it might be a false floor
and he could fall through.

Then he noticed an enormous cone had built up from snow sift-
ing in from outside. It was Per Bak's sand pile. The cone started at
this deep level and went straight up, narrowing to the hole in the
ceiling. Simpson had been unable to see the formation from his
previous position above, and anyway, it had been dark. Now he
realized the sun had risen. The cone was very steep, but he decided
that he could climb it. Besides, there was no choice. That's the
whole point about survival: No one would be mad enough to do
these things if it weren't absolutely necessary.

After that harrowing descent, his brain needed a rest. The
organism knows what it wants and finds what it needs. Still hanging on his rope, Simpson began to experience a sense of wonder
and even joy at his environment, that same spiritual and mystical
transformation reported by Steve Callahan and many other survivors. It is always followed by a certainty of survival and a

renewed commitment. With dawn came light, and with light came revelation:

> A pillar of gold light beamed diagonally from a small hole in the roof, spraying bright reflections off the far wall of the crevasse. I was mesmerized by this beam of sunlight burning through the vaulted ceiling from the real world outside. It had me so fixated that I forgot about the uncertain floor below and let myself slide down the rest of the rope. I was going to reach that sunbeam. I knew it then with absolute certainty. How I would do it, and when I would reach it were not considered. I just knew.

Lauren Elder and Debbie Kiley both reported similar experiences. Indeed, Christopher Burney's concentration camp experience of seeing summer come was like that—spiritual wonder, followed by commitment.

Survivors always turn a bad situation into an advantage or at least an opportunity. As Joe Simpson reported: "In seconds my whole outlook changed. . . . I could do something positive. I could crawl and climb. . . . Now I had a plan." Here he was in hell, and he had just glimpsed redemption. "The change in me was astonishing. I felt invigorated, full of energy and optimism." Again, he laughed out loud. "I could see possible dangers, very real risks that could destroy my hopes, but somehow I knew I could overcome them." Good survivors, like good wives, husbands, and CEOs, always consider the bleak side of things, too. They plan for them and have an earnest hope that they will manage. But they do not care overly much that they might not. They accept that to succumb is always a possibility and is ultimately their fate. They know safety is an illusion and being obsessed with safety is a sickness. They have a frank relationship with risk, which is the essence of life. They don't need others to take care of them. They are used to caring for themselves and facing the inherent hazards of life. So when something big happens, when they are in deep trouble, it is just more of the same, and they proceed in more or less the same way: They endure.

Simpson knew where he was and where he was going. He fixed a goal. What else do we ever do? Survival is nothing more than an ordinary life well lived in extreme circumstances. He embraced the world, and the world embraced him. He was not lost, so all was not lost. And as we learned from Ken Killip, not being lost is not a matter of getting back to where you started from; it is a decision not to be lost whereever you happen to find yourself. It's simply saying, "I'm not lost, I'm right here." Which is, significantly, what a child would say. Zen mind. Beginner's mind.

As Simpson prepared to climb, to project himself into a future that he had made as real as the past, he took in his environment, drew in its strength and beauty once more. "For all its hushed cold menace, there was a feeling of sacredness about the chamber, with its magnificent vaulted crystal ceiling, its gleaming walls encrusted with . . . myriad fallen stones. . . . It had me now, and without the sunbeams I might have sat there numbed and defeated." But, of course, the sunbeams are always there. The trick is in seeing them.

At such times, facing their biggest challenge, survivors also find that they even feel lucky. After catching a fish, Steve Callahan wrote: "I sit a thousand miles away from any companionship, money, or luxury, yet I have a feeling of wealth." Joe Simpson said that he "kept telling myself that I was lucky to have found a slope at all."

Saint-Exupéry's plane went down in the Lybian Desert, a circumstance at least as grave as Simpson's. He wrote in *Airman's Odyssey*: "Here we are, condemned to death, and still the certainty of dying cannot compare with the pleasure I am feeling. The joy I take from this half an orange which I am holding in my hand is one of the greatest joys I have ever known." At no time did he stop to bemoan his fate, or if he did, it was only to laugh at himself.

So Simpson returned to his old friend, the patterns and rhythms of the dance. He'd place his ice tools, plant his good foot, then pull hard while hopping upward to plant the foot a bit higher up. "Bend, hop, rest; bend, hop, rest . . ." He found once more that concentrating on the pattern helped him ignore the pain.

At the same time that he took the greatest risk, he was also safeguarding himself. "I resisted the urge to look up or down. I knew that I was making desperately slow progress and I didn't want to be reminded of it by seeing the sunbeam still far above me." Interesting. For how can you know something and still keep from reminding yourself of it? It would seem to be a paradox . . . unless you had two brains. Indeed, when Simpson says he "knew," he means that his hippocampus contained a short-term memory pattern about his situation. But the visual perception of how far he had to go would be sent straight down through the thalamus to the amygdala to be screened, and that was what he wished to avoid, because it might trigger a whole new set of emotions that he didn't need right now. He could bump what was in working memory out by concentrating on something else: his pattern. It was a wise choice. Keep the perceptual input to a minimum right now. Don't feed that amygdala any scary raw data.

SIMPSON HAD been in the crevasse for twelve hours when he stuck his head out of the hole and into the sunlight to see "beauty I had never noticed before." Once fully out, he collapsed in the snow, releasing his tight grip on himself at last. That lock he had on himself could be relaxed for a little while. Not too long, but he deserved a rest. It was only then that Simpson realized what he already knew: He was still very high on the mountain, six miles above the campsite and help. And he had a broken leg. He was exhausted, dehydrated, hypothermic, frostbitten, and had just used his last ice screw. Since he was officially "dead," no one would come looking for him. How the hell, he began to wonder, would he get out? All at once, it seemed that there was some sort of dark malevolent force to this place, dogging him personally.

In *Lord Jim*, Joseph Conrad describes "these elemental furies . . . coming at him with a purpose of malice . . . which means to sweep the whole precious world utterly away from his sight by the simple and appalling fact of taking his life."

Simpson had protected himself from such thoughts while climbing out of the crevasse. If he'd thought of death, he might not have bothered. Like walling off Per Bak's sand pile against collapse, he'd compartmentalized and set his short-term goal, not even that of climbing the cone but of doing each pattern meticulously, one step at a time. Thus he had been able to go boldly into the unknown. Now he faced a new unknown: somehow to cover the impossibly rugged six miles to camp. His situation seemed every bit as hopeless as it had been down in the hole. That's why a survivor must compartmentalize and set small goals. It would kill him not to.

Simpson "felt crushed" as he realized that the movie had to start over—the friggin' movie!—right from the first stupid clichéd reel. It was enough to make you do what all good survivors do: He laughed at himself. "This is getting ridiculous," he said out loud. You become like a slapstick version of yourself, taking pratfalls and suffering impossible abuses, like Laurel and Hardy getting run over by a steamroller and getting back up again. That's why we love to see comic characters suffer. Charlie Chaplin is survival training, after all. To laugh at our own misfortune, we must be willing to play the fool. It keeps us from taking ourselves too seriously. It keeps us humble.

Simpson took a last look around and accepted that he would almost certainly die out here. From long habit, he was able to view his own death dispassionately; but he wouldn't die without a struggle. He had something left to risk, so he would risk it. A survivor builds up an account of commitment over a lifetime. The more he invests, the more he has when trouble comes. Now Simpson felt his senses grow "sharp and alert."

Like Callahan, he saw himself, humbly, with spiritual empathy, and even gratitude, as a part of this sweet old world. He looked across the immense broken landscape he had to cross, looked "ahead at the land stretching into distant haze and saw my part in it with a greater clarity and honesty than I had ever experienced before. . . . An excited tingle ran down my spine. I was committed. The game had taken over." He felt honored to be in a tournament with the greatest chess player of all.

With the practice he'd had during the past few crises (and through the years as an extreme mountaineer), Simpson was able to reconsolidate rapidly. He went through all the stages and organized his attitudes and actions. His initial horror, anger, and depression at realizing how far he was from success rapidly turned into a will and a plan. His mind split once again. His perceptions became sharp as he accepted his world and embraced the immediate tasks ahead of him. He was back inside himself. "There were no dark forces acting against me," he realized. "A voice in my head told me that this was true, cutting through the jumble in my mind with its coldly rational sound." He listened to the rational one, which directed correct action, while "The other mind rambled out a disconnected series of images, and memories and hopes, which I attended to in a day dream state as I set about obeying the orders of the *voice*." He fashioned a splint and used his ice ax as a cane. The patterns took over once more: "Place-lift-brace-hop. Place-lift-brace-hop."

He could actually hear the rational voice, while the other was just a tangle of feelings. The neocortex has connections to the auditory thalamus, too, and if you can make it overpower the amygdala and tone down the hippocampal memories, then that's what you hear, a rational voice, "clean and sharp and commanding."

Every survivor hears the voice. Remarque wrote, "A clear voice utters words that bring me peace, to me, a soldier in big boots."

Yet despite the noble struggle and the correct actions, despite his great attitude, Joe Simpson was dying. Nature's forces are effectively limitless; human energy is not. The various stresses and assaults had overwhelmed his thinking; gradually, he began to lose that fine balance between the emotional drive to action and his overall cognitive control.

Simpson began wandering. He pressed on longer than he should have that night. His energy was getting dangerously low. Then he became lost and no longer knew the way. Now he began going through the stages of getting lost, which were for him, at last, the stages of dying. By chance on his meandering path, he blundered into a snow slope. The far-off voice spoke up at last, and Simpson dug a snow cave for the night and gave in to fatigue.

He awoke in the morning quoting Shakespeare to himself, shouting at the roof of his snow cave. "I felt delighted . . . bellowing the words in my best Laurence Olivier voice. . . ." When the brain is allowed to relax its grip on the survival struggle for a moment, as we've seen, it eagerly gets to work. And where there is no new input to process, to map or memorize, it jumps up and gives you a core dump. That's when you'd better hope you've spent your life building a core.

But Simpson was still in grave danger. He'd exhausted himself badly the night before. Fatigue is mysterious and profound; you don't just sleep it off. He needed water, food, and advanced life support for his injuries.

As he hobbled on, he made an effort to find his patterns again. But pain and thirst were now tormenting him. He had worked out a system of going from rock to rock, timing himself between those short-term goals, to keep working memory occupied. But his resistance was dwindling. Thoughts of the flowing water at a stream he'd seen on the ascent preyed on his mind. The logical voice faded even as that other, more disorganized voice urged him on: *Go now! Hurry! Water is just ahead!*

Simpson stopped timing his moves and scrambled on, just to get to water as fast as possible.

Night fell again, and he still had not found water. Another whole day had passed. Then his headlamp failed. Hobbling on the makeshift splint, he fell and could not get up. He lay on the rocks where he'd collapsed and slept only fitfully.

It was as if the cognitive part of his brain, once freed from the effort of control, got busy doing what it normally did. "Now that I was safe to sleep I felt sharply awake." All the fear he had resisted came flooding in once those emotional parts of his brain were freed from their leash. The pain in his leg, which he'd been suppressing, surged back into consciousness to torment him. All the worries that he'd kept at bay now plagued him.

The next day, "I had hit some sort of wall." His energy was gone. His will was slipping away. He lay idling, unable to move; but then he thought of Yates and remembered that Yates believed he was dead. There would be no reason for him to stay on the mountain. He'd leave their camp. Even if Simpson managed to get there, he'd die. The certainty of death suddenly energized Simpson. "I must reach camp today," he told himself.

At that moment, Yates was burning Simpson's clothes and packing up to return home.

Simpson knew that he had ten hours of daylight left. He recognized, as Steve Callahan had, the need to do everything perfectly. He began "to break those hours into short stages, each one carefully timed." Once more, using his anger as a spur, he forced himself into a rigorous pattern, rhythm, and schedule of motion across terrain that would have given any two-legged hiker trouble.

He almost didn't make it for any number of reasons. Never underestimate the need for an adult portion of luck. "If you get a lucky break you have to really use it," he later told *Backpacker Magazine*. "You have to fight like a bastard. You can't just sit there and wait to get lucky. It doesn't happen." Yates was about to leave;

the locals were bringing up the burros. Simpson was succumbing to dehydration and fatigue as he literally flopped around on the rocks just beyond their camp. At last he managed to seize control of his parched throat and cry out. Yates heard the cry but did not believe it possible. He went to investigate. Simpson was taken out of the Peruvian Andes on a donkey. He's been on a dozen Himalayan climbs since then and the route that Simpson and Yates took up Siula Grande has never been repeated.

FOURTEEN

A CERTAIN
NOBILITY

ONE DAY I was chilling in Beaver Flight at Randolph Air Force Base during a weather standby. We couldn't fly because of a thunderstorm. I was idling around, reading a sign on the bulletin board that said: I WANT TO DIE IN MY SLEEP LIKE MY GRANDFATHER. And then, in smaller letters, *Not yelling and screaming like the passengers in his car.* Typical fighter pilot humor. (Their group was technically known as Bravo Flight, Bravo being the semaphore designator for the letter B.)

I heard a pilot named Chuck come in, a fun guy, buzz cut with a square head and shoulders, and like the rest of us, not too tall. Chuck had left Iceland and his F-15 Eagle to become an instructor. He told me about how he and the other Air Force pilots would play chicken at night with the Russians. Then these big Russian ships would come along with multimillion-candlepower lights to blind them. "We couldn't see a fucking thing for like half an hour after that," he told me. It was all a game. A game of nerves. There was no shooting allowed, which presumably would have triggered World War III, but you could die just by losing your cool.

It was nasty work. The glory of the sky is one thing, but those guys had to sit for endless hours, most of them uneventful, pissing through a tube into their "piddle pack" with a rubber mask over their faces (or not). The temperature in the tiny cockpit was either way too hot or way too cold. Food was out of the question except perhaps for a stolen snack. I'd made some long flights in fighter aircraft and knew how heavy that helmet got, too. We wore a silk skullcap, but it did little good. After an hour, it felt like a hot iron was sitting in the center of my skull, and the weight of it gradually compressed the discs in my neck until the ache went all the way down my back. The helmet and face mask have to be specially fitted, so before I flew, one of the life-support crew measured my face with calipers from the bridge of my nose to the tip of my chin. They must be tight, because the wicked wind you meet upon ejecting can blow them right off if they're not. So your head and face are in a vise, too.

Nomex gloves, jump boots, Nomex clothing, including a closed collar to keep the flames from burning exposed skin. It was like being swathed in Saran Wrap. And there was only one position the body could be in, because feet, hands, and head were all required to fly. (Yes, you could use autopilot and you could even get buck naked over Kuwait, but few did.) After a couple of hours, even your teeth hurt. And that's before anything bad happened.

Everybody wanted to get into F-15s, but everybody wanted to go home after a while, too. I certainly understood why Chuck had come here as a student IP. The job of the Twelfth Flying Training Wing at Randolph was to turn experienced fighter pilots into instructors to train other fighter pilots. There would be short flights, long lazy days of briefings in the classroom, and playing Krud (pool played without cues or rules) with the other guys at Friday night beer call in the Officers' Club. It was a pretty cushy job, although, as my instructor, Captain Stuart Rodgers, told me, "We're teaching these trainees the only form of combat flying where the enemy is going to be in the cockpit with you." What he meant was that the instructor of the IP trainee would pretend to be a new student pilot, while the trainee pretended to be his instructor. And the pretend student pilot would try to do something so stupid that it would kill them both, while the trainee tried to anticipate his moves and stop him before it got serious.

I'd just asked Chuck about the pay, and he turned from our conversation and shouted out to the room in general: "Anybody who's doing this for the money, raise your hand." Everyone laughed. Chuck swore that the guy who drove us across the ramp to the airplanes in a blue van every morning made more money than the pilots.

Chuck had been doing dissimilar air combat for three years. He and three other Eagle drivers would take off and fly out over the ocean, not knowing what they'd meet. Somewhere out there they'd run into a bunch of Navy guys whose job it was to simulate bandits and bogies, Soviets at the time. Some even had a few real MiGs, too. When the dogfight started, with planes closing at supersonic speeds, all of the aircraft turned in on one another at once. Everyone was trying to get on someone else's tail, creating a mess they called a furball. On that rainy day in Beaver Flight, we sat watching a video of two planes in just such a simulated dogfight. The pilots in B Flight gathered around the TV and watched the sequence over and over, listening carefully to the radio transmissions, which told the story of what those pilots did in an emergency of amazing proportions, a midair collision in a dogfight.

The first trans was the leader, calling: "Knock it off! Knock it off! Knock it off!" which was the standard order to disengage from the furball. (Emergency transmissions are always sent three times, as in "Mayday, Mayday, Mayday.")

We saw the two planes on the screen glide past each other, shearing off body parts with splashes of smoke like the strokes of a Japanese sumi-e brush. Then one of the pilots who was hit called: "Bailout! Bailout! Bailout!" And a moment later: "Mark."

Then—*Bang*!—his seat fired in a plume of smoke, and he was out.

"That's just pure polish," Chuck told me. "In a situation like that, your only responsibility is to get out. But he said, 'Mark,' so they'd mark his position when he ejected, to make it easier to find him. That's just pure professionalism."

Pure professionalism. That's what survival is all about. At the core of that pilot's professionalism was the fact that he thought of his teammates in the direst of emergencies. He was not a victim. Madeline Amy Sweeney, a flight attendant on American Airlines Flight 11 on September 11, 2001, demonstrated the same sort of pure professionalism as she sat in a rear seat and calmly reported over the phone what was going on inside the plane that was about to hit the north tower of the World Trade Center. As her flight service manager heard her describe a passenger having his throat slit, the position of the hijackers, and other important information that would never have gotten out otherwise, he could not help but be impressed by her remarkable control.

Like the fighter pilot who ejected, she was what Peter Leschak calls, "a player—a fighter and helper from beginning to end." It is one thing to say it, another to do it, and yet another still to do it in the clutch of mortal risk. Ejecting from an airplane is not a trivial matter. It is a violent, dangerous, and catastrophic maneuver, like sitting on a stick of dynamite. It often breaks the pilot's back and puts him permanently out of the service. It can promptly remove elbows, knees, and other body parts; and if not done correctly, it can kill you. The pilots know that and are often more afraid to eject than to ride a crippled plane down to a crash landing. Therefore, when they reach down (or up) to pull the handle, it feels a bit like placing a gun to your own head and preparing to pull the trigger. This pilot, whose plane was catastrophically disabled, knew that he must either crash and die or else pull the handle and risk landing in the sea critically injured with the attendant possibility of drowning. In that terrible moment, he thought of his crew and gave them the one thing they needed to avoid getting themselves hurt while trying to save him: his position.

In his neurological researches, Joseph LeDoux wrote, "One of the reasons that cognition is so useful a part of the mental arsenal is that it allows this shift from *reaction* to *action*. The survival advantages that come from being able to make this shift may have been an important ingredient that shaped the evolutionary elaboration of cognition in mammals and the explosion of cognition in primates, especially humans." When Neanderthals and the weaker *Homo sapiens* found themselves in the same environment 50,000 or so years ago, Neanderthals went extinct, but we survived. And that higher cognition may have made all the difference.

In the complex technological world we've created, cognition, problem solving, becomes more and more important for our continued existence, so we come to depend on it as if it were infallible. But that strategy often gives an illusory sense of control that fails us in moments of crisis. Its stepwise linearity is no match for nonlinear dynamic systems, which can behave in turbulent, unpredictable ways, making quantum leaps as they rapidly change their nature. At the moment of truth that defines survivors, and if reason is the only tool at your disposal, you are likely to achieve what David Ruelle, a Belgian physicist and one of the founders of chaos theory, referred to as "the steady state of death."

IN HERODOTUS' *Histories*, Solon of Athens visits Croesus, who shows him his vast treasuries and then says, "My dear guest from Athens, we have often heard about you in Sardis: You are famous for your learning and your travels. We hear that you love knowledge and have journeyed far and wide, to see the world. So I really want to ask you whether you have ever come across anyone who is happier than everyone else?"

Croesus expects to be nominated for the position, but Solon instead names an ordinary man, Tellus of Athens. "You see," Solon explains to the shocked king, "in a battle at Eleusis between Athens and her neighbors he stepped into the breach and made the enemy turn tail and flee; he died, but his death was splendid, and the Athenians awarded him a public funeral on the spot where he fell, and greatly honored him."

When Croesus grows angry at not being named the happiest man in the world, Solon counsels him:

> I won't be in a position to say what you're asking me to say about you until I find out that you died well. You see, someone with vast wealth is no better off than someone who lives from day to day, unless good fortune attends him and sees to it that, when he dies, he dies well and with all his advantages intact. . . . It is necessary to consider the end of anything, however, and to see how it will turn out, because the god often offers prosperity to men, but then destroys them utterly and completely.

So it is with the paradox of survival. There is no one point at which a man or woman can be said to have survived and be a perfect survivor until he or she is dead, for each test is a preparation for the next. By its very nature, a survival situation, like life itself, comes upon you unexpectedly. So only if a life is lived in its entirety as an act of survival can you have a hope of finding correct action at the moment of crisis. Survival is a path that must be walked from birth to death. It is a way of life. Asked to judge a life, Solon chose to judge one that was complete.

Ancient books of wisdom often treat issues that bear directly on survival. The *Tao Te Ching* is filled with glorious paradoxes and compelling mysteries. ("The ways that can be walked are not the eternal Way;/ The names that can be named are not the eternal name.") Plunging into it with the expectation that you will be told what to do is a frustrating experience. The book is meant to lead to revelation, but the path is long. As St. Paul the apostle wrote, you must "work out your own salvation with fear and trembling." There is instant wilderness but no instant survival.

The *Tao Te Ching* does not contain salvation, but the Way can be followed with integrity. The text explains survival in this way:

One who is good at preserving life
    does not avoid tigers and rhinoceroses
        when he walks in the hills;
    nor does he put on armor and take up weapons
        when he enters battle.
    The rhinoceros has no place to jab its horn,
    The tiger has no place to fasten its claws,
    Weapons have no place to admit their blades.
Now,
    What is the reason for this?
    Because on him there are no mortal spots.

How to achieve that state is the key to and the central mystery of survival. To be, as Peter Leschak put it, "so alive you simply cannot die." He has survived "twenty-one seasons—560 fires in 11 states and one Canadian province—because I've tapped into the spiritual aspects of working fire." That is not what you'd expect to be told in training nor to hear from a tough, blue-collar firefighting helicopter boss who works his rough crews on hellacious wilderness fires. "There is a core of mystery and faith that has guided not only my career but also my life," he writes. "To me, the fireground is a sacred locale, a place of power that is rich not only in tradition, history, and ecological imperatives, but also in sources of emotion and meditations that I can only describe in terms of reverence and awe." Leschak's sense of wealth and wonder in the face of a calamity that could easily include his own death is typical of the best survivors. To revere pain and fear, to embrace friction, are bedrock skills of survival. Leschak, who left behind his schooling for the ministry in order to risk his life routinely as a firefighter, reflects: "I've become a minister after all. . . . I've struck a wellspring of spiritual inspiration and practice."

One of the reasons that astronauts are so good at what they do and have so few fatal accidents is that they, too, have struck that wellspring. They are ministers. They are cloistered at Johnson Space Center in Houston for their training. They lead ascetic lives of denial and service. They learn to function as a team, to value their brothers and sisters above their own lives and to protect them. They learn to become intimate with fear and pain, to make fear and pain tools of their salvation. They train for years and years, knowing that they may never even get the chance to fly. Like the student who wanted to become a Zen swordsman, they are forced to toil and endure the blows of their masters until they no longer need swordsmanship. Only then are they worthy to hold the sword. The selection process at NASA seeks out those who are already on that path when they arrive. Whatever an astronaut is, he is not made that way by NASA. He arrives with it in his heart, and the purpose of his training is to extend and deepen those transcendent qualities.

When Neil Armstrong landed on the moon, he had to take control away from *Eagle* and fly by the seat of his pants. The expression isn't gratuitous. It means that he had to engage his body (read: emotions) as well as his thinking with the machine (read: the environment). In the ministry of NASA, he had trained his emotions as well as his reason. Thinking too much or too little would not do. Here was a moment of truth in which correct action would arise only out of a perfect blending of the linear and nonlinear, reason and emotion (those instant emotional bookmarks).

The proof that logic could not lead to correct action was that *Eagle* had been set up as an automated, logical system. The computer had a program to follow all the way to the surface of the moon. Armstrong had trained assiduously to land it by hand, but the plan was that *Eagle* would land itself and he would simply monitor the systems. When Armstrong saw that *Eagle* was going to land in a boulder field, he knew that logic had failed at the critical moment. He took the controls.

As he piloted *Eagle* down toward the surface of the moon,

numerous things went wrong and alarms were blaring in his ears. When he approached to within 1,000 feet of the surface, he saw the area where he intended to land—"A crater as big as a football field was just ahead, surrounded by a field of boulders, some as big as Volkswagens," according to Andrew Chaikin's *A Man on the Moon*. Even as he flew past that spot, using precious fuel, Armstrong found no place to land. His heart rate had gone from 77 to 156 beats a minute, clearly showing that his emotional systems were fully engaged. But his voice was calm.

By logic, he should have aborted the mission. There wasn't enough fuel and there wasn't enough time. But logic knows nothing of glory. He was using the spiritual bounty from the ministry of his long training, and the output flowed as rapid-fire bursts of energy from numerous neural networks, animating dynamic models of his world with emotional bookmarks and rational thoughts in ever-changing combinations to keep Armstrong's feet, hands, and eyes moving in such a way as to keep *Eagle* ahead of the changes in the dynamic system involving mass, momentum, speed, gravity, burn rate, and the fast-approaching lunar surface. The event was literally flowing through him and out of him. He was processing the world and adapting to it at a fantastic rate. And he was doing it in what Leschak calls a "tenaciously unforgiving environment," which is what made his performance so remarkable. At the same time that he was engaged with that ineffable flow of survival actions, the ballast of his reason held fast, and he was able to quiet the alarm bells, both from the ship and from inside his emotional system. Neurologically speaking, it was an Olympian performance.

He chose another area to land, but it was no good: more boulders. Zooming along, he chose yet another. That was no good either. At last he saw a decent spot. With only ninety seconds of fuel left (twenty of which he needed for an abort), he headed for it just 100 feet off the surface. As he began at last to settle to a touchdown, *Eagle*'s rocket engines started blowing the lunar dust around, putting him in a gray-out.

Armstrong let the visual cues drop away without wavering on the controls. The gravity of the moon was so faint that he didn't even have vestibular information—the seat of the pants—as he gentled *Eagle* to the ground. He even had the presence of mind to assess his own behavior, as many elite performers do in the clinch. As Chaikin recounts, "He was displeased with himself, sure that he was not flying *Eagle* smoothly." He had twenty seconds worth of fuel left when he shut the engine down, precisely the amount he needed to get home.

Leschak's description of his twenty years of risking his life as a wildland firefighter applies exactly: "I'm a secular priest ordained by training, experience, and, most important, the willingness to accept the mantle of command. That willingness encompasses the realization that failure is easy, and such failure could kill me or, worse, kill someone else." And as was true of the pilots in Beaver Flight, "It's the nature of our society that I'm considered unworthy of huge financial reward for that risk. But what can be earned is a certain nobility—not in the sense of aristocratic status but in the sense of striving for quality and dignity of behavior and living."

IN 1988, *Harper's* published an article I wrote about flying with the IPs of Beaver Flight, as they humorously called themselves. At the time, I was hot on the quest to enter my father's realm of cool and heroism, and so I had no idea that I was talking about the essentials of survival when I wrote of their humor and their modest cool. I asked Captain Rodgers, my instructor, how they selected candidates to be Air Force fighter pilots. Like Croesus pestering Solon, I was perhaps secretly hoping that he'd say that I was the ideal candidate. Instead, he said, "Well, you see, we go down to the Dunkin' Donuts, and we take everybody who's sitting at the counter, and we say, 'Okay, you're pilots.' Then we send them out to an airfield somewhere real far away, and we put them in airplanes, and see if they can fly."

In a more serious frame of mind one day, he told me what makes a good fighter pilot. "He's the kind of person who, when you tell him something, he gets it right away. He's a good athlete. He's competitive. He can stand up in front of a hundred people and speak on something he doesn't know about. He'll just do it. He won't hesitate. He's a creature of habit and a perfectionist." The pilots who are selected to fly fighter planes are chosen for their academic performance as well as their flying grades. They're taken from the top 5 percent, and the grades have to be calculated to two decimal places just to establish the rankings, he told me. He was describing my father. He was describing the perfect survivor, too.

The image of the fighter pilot (or the astronaut, for that matter) that has been created by popular movies and literature was true in my father's day, but not any more. In the urgency of war, the entire process was different, and in a way it was necessary for the Army to enlist a sort of airborne Hell's Angels just to get the job done. (My father described the physical exam this way: The doctor said, "Can you see lightning and hear thunder? Congratulations, you're now a pilot.") In the midst of the Cold War, when President Kennedy promised to put a man on the moon by the end of the decade, the same urgency applied to sending men into space. The archetype described in *The Right Stuff* was true then, but it's not true now. There were pilots in my father's day who would drink all night and were rumored to put tobacco in their eyes just to keep them open while they flew the next day's mission. One was said to bungee the stick to the window frame and sleep on the way into battle. I can't say for sure if that's true, but knowing the people I've met, it wouldn't surprise me. The astronaut of today is more likely to have a Ph.D. than a drinking problem.

The modern fighter pilot does not strut around grabbing fannies. For one thing, some of the pilots are women, and the Navy's experience with grabbing fannies demonstrated what the consequences can be. The Friday-night Krud games can get violent and people do get drunk, but that's one night a week, and anybody who drinks much more often finds himself flying a desk. No one buzzes the tower and stays in the Air Force, either.

I saw some pilots shooting down their watches and slapping high-fives on the ramp one day and mistakenly thought they really were like Tom Cruise in *Top Gun*. As I approached the van for the ride out to the airplane, I heard one of them say, "I feel the need," and another one answer, "The need for speed!"

Captain Rodgers explained to me that they were making fun of the movie. "If a pilot here acted like Tom Cruise acted in that movie, he would have no friends. He'd be an outcast." The same arrogance and reckless disregard for safety that will kill you in a high-performance aircraft will kill you in the wilderness or in any other place of high objective hazard. The modern craze for high-risk sports and wilderness recreation, which has been embraced by so many naive travelers, is bound to result in accidents, but the images such myths project make it even more likely. An ad for a Nissan SUV, for example, showed a mountain biker through the open hatchback of the truck. He's injured, his limbs in makeshift traction, and his bike has been destroyed. The headline read: RECOVERY ROOM. The implication is that part of the fun is in getting hurt. But hurting and getting hurt are two different things.

A Zen priest I know studied to be an archer. He described his first lesson to me. The master took him to a cliff at dawn. He set up a target and strung his bow. Then he drew an arrow and aimed at the target. With the bow taut, he turned at the last moment and shot the arrow out over the cliff.

"We are officers first," Captain Rodgers explained, "pilots second." The flying is not the point, just as hitting the target is not the point of the Zen discipline of archery. They are both side effects of a way of life. Side effects of integrity that lead to the Way. Once you no longer need or desire to hit the target, then the Way leads you there. Being one with the bow, you do not have to hit the target. The bow does it for you. Integrity begins with humility. When I was finishing my training, Captain Rodgers received word that he'd been assigned to F-15s, a coveted job. I asked him whether it was because he was the best. "No, no," he laughed, embarrassed by the suggestion. "I was just lucky. Real lucky."

I thought his choice of words was telling. Once my father got his airplane shot to pieces and was decorated for saving his crew. Years later, when I asked him how he'd done it, he said, "I was just lucky. Very lucky." And he laughed.

FIFTEEN

THE DAY OF
THE FALL

THAT VECTOR LEADING to survival, which Joe Simpson and Steve Callahan took, stretches back into childhood. To enter the wilderness, to challenge the forces of nature, we must be worthy, and worthiness doesn't come from a weekend survival school, the Eagle Scouts, or even a few years in the military. Peter Leschak wrote, "In fire and other emergency operations, you must not merely tolerate uncertainty, you must savor it. Or you won't last long. The most efficient preparation is a general mental, physical, and professional readiness nurtured over years of training and experience. You live to live. Preparing is itself an activity, and action is preparation." He's talking about making himself worthy of survival, and his way of doing it in the wilderness is with the added burden of fire, just as my father's manner of flying, itself an act of survival, was to do it while people were shooting at him.

I first learned about being worthy from my father. I learned again when I became a pilot. And again when I became an instrument pilot, a commercial pilot, and then an aerobatics pilot. Flying bush pilot planes in the Arctic regions of Alaska—the Brooks

Range and on up the coast past Wainwright to Barrow—I learned, too, about indifferent forces that punish inattention or arrogance. When I was competing with the International Aerobatics Club, even as I saw those around me being killed, I realized that I had to be at once bold and humble, that I had to open my mind to this energetic world, which never sits still, the complex churning of its materials, from which I'd made my own Braille language of life.

My father was too badly injured in his crash to continue as a pilot and went back to school to become a medical school professor, a scientist. I followed him to the University of Texas, then to Baylor Medical School and at last to Northwestern, and grew up working in his labs, eventually operating an electron microscope and peering with him into the very machinery of human cells. I'd go to his classes so that I'd be able to speak his language, the language of science. When he took the podium, he always began by saying, "Fellow students . . ." He taught me the humility of knowing that we were all, always, students, and that to stop being a student was to stop living.

When he turned seventy years old, I was hot and heavy on the contest circuit with the International Aerobatics Club. I took him up for his birthday to show him my routine of spins, loops, rolls, hammerheads, Cuban eights, Immelmans, and split-S's, a continuous corkscrewing of the airplane, which let one maneuver lead into the next in a sort of high-octane gasoline ballet.

A plane is a noisy, stinking thing to those on the ground, but to the pilot it can sometimes seem absolutely silent, like a sailboat (until you hear the wrong sound, and then it gets your prompt attention). On that flight, my father sat quietly in the tandem seat as I ripped the plane through four and five G's, climbing, descending, rolling, and falling through the hard air, switching blue sky for green earth a dozen times a minute as the smooth beauty of the whirling world filled me with wonder and joy. It didn't feel as if I flew the plane. It felt as if I'd become the plane; the wingtips had nerves.

My brother Michael, one of my father's students and a physician, had expressed some concern that the G-forces might be bad for our father at his age. But when I was done and my wheels barked onto the asphalt, my father climbed out and said, "You're a really good pilot." He did not give praise idly. It was one of the most important moments in my life. I was worthy. Air worthy.

WHEN PEOPLE hear about my father's survival, they think of the long fall from the sky or the moment when that German peasant, standing on the stub of wing, pulled the trigger on his old pistol and it misfired. But those singular events are not the point.

Sure, it takes luck to be a survivor, and luck is nothing more than the accumulation of circumstance throughout a life. One year, I arrived in Glacier National Park to watch the biggest snow clearing operation in the United States. The big bend near the apex of Going To The Sun Road can be 100 feet deep in snow, and the road is only two lanes wide. Avalanches regularly rip through there, sometimes sweeping men and machines off into the couloirs. As the weather warms, the cliffs calve rocks the size of automobiles. As I settled in with the crew, the snow boss told me that the previous season, on the day the road opened, a 30-ton rock had fallen onto a car, killing a Japanese tourist while sparing his wife in the passenger seat next to him. And I thought: All his life he drove along roads to get to that exact spot at that exact moment. And so did his wife, who survived. But her survival didn't end at that moment, it began there. Her task was to survive the terrible event, to go on and live her life. So with my father. The lesson of survival that I took from his story was not that he was so lucky as to fall 27,000 feet and not die. It was that he had to have the strength to go on and live sixty more years after losing his beloved brothers, his crew, after breaking his body into so many pieces, after prison camp. He was twenty-three years old and had to forge a strategy for surviving everything else. I'd seen many of his fellow combatants simply give up, collapsed old men, walking ghosts. That the German peasant's old and badly abused pistol jammed was sheer chance. Everything after that was not.

As he lay there in a heap by the rudder pedals, my father watched his would-be assassin with a sort of dim, swooning amusement as the man tried to get the firing mechanism sorted out. Then my father began laughing, which infuriated the German, who was cursing a blue streak. My father was able to understand German reasonably well and was struck by the movie-like quality of the scene. It was all a bit much: to get blown out of the sky and fall 27,000 feet without a parachute—and survive—only to land in the exact spot where there's a pissed-off farmer with a gun. He couldn't stop laughing. It was the beginning of his salvation, not the end. Humor was the key.

A German officer appeared and told the farmer that he could not shoot the American pilot, who was officially a prisoner of the German Reich. There was an argument. Harsh words. The peasant said that the pilot deserved to die for bombing them, and anyway, he wasn't going to live long. Look at him. Indeed, his nose had been cut off, he was bleeding profusely, and he was crumpled in a bloody, mangled heap. He was obviously delirious. Look, he's laughing.

While they were arguing about his fate, Mrs. Peiffer came out from her farmhouse outside the town of Neuss (now a suburb of Düsseldorf). The front half of the B-17 had come down on the side of a railroad embankment that bordered her land, and she was hopping mad. (The aft portion of the plane had crashed about half a mile away with some of the crew, one of whom had lost his legs somewhere in the sky.) She'd seen the whole thing from her house. For some time now she had refused to take shelter against the air raids. The German soldiers were all young, and the woman took advantage of her age, ordering them to care for the wounded pilot.

. . .

MY FATHER awoke in the snow, laid out with some of his dead crew. "I was in and out of consciousness," he told me. "But I was deliriously happy. Maybe it was because of my injuries. Maybe someone had given me morphine. I don't know. But I felt no pain, and I was just happy to be alive."

But to his left was Colonel Hunter, his commandant and co-pilot for the day. My father was captain of the ship, and as such, he was responsible for the safety of all on board. Now Hunter lay dead in the newly fallen snow, and the lieutenant couldn't help feeling guilty about how happy he was to be alive when all the rest were dead.

While he struggled with the confusing emotions, he began vomiting blood. He concluded that he must have internal injuries. Suddenly, his joy turned to terror as he realized that he was going to die. After all that, to perish in the snow. He began crying, and a German soldier, himself no more than a boy, came over to see what the trouble was. He reached down and flipped the American boy's nose back into place for him. Although it had been cut off by flying glass or metal, it had been hanging by a flap of skin, and now my father understood: He'd been lying on his back, swallowing all the blood from his flesh wound. That's why he was throwing up. Once more, he was overcome with joy: He was going to live!

He passed out again.

Mrs. Peiffer ordered the German soldiers to carry the wounded American lieutenant into her house, and when he awoke the next time, they had laid him before her fireplace. She gave him tea and a cigarette. As both his arms, both hands, both feet, both legs, and numerous ribs were broken, she had to hold the tea and help him smoke his cigarette. And he thought: This isn't going to be so bad. Maybe this is what German prison camp is like, tea and cigarettes before a cozy fire.

Then a truck was pulled up to the house and he was thrown into the back of it and driven overland. "As soon as we started bouncing across that frozen ground," he said, "I could feel the broken bones grinding against each other." The pain was so excruciating that he couldn't stop screaming until he mercifully passed out once more. But each time he came to, he awoke screaming.

At last they arrived at the prison camp near Gerresheim, where he was thrown in a basement with prisoners from all over Europe and America. By chance, one of them, Dr. Géri, was a member of the French Resistance. He was a surgeon and had been allowed some meager medical supplies with which to treat the wounded. There were also a few male nurses who were allowed to work in the crude lazaret.

Dr. Géri wired my father up with piano wire and plastered him all over until he looked like a great albino spider hanging from the basement ceiling beneath a single bare globe, which was strung on a length of electric cord.

In the ensuing days and weeks, Dr. Géri would have to tighten the wires—to tune the piano—and my father would scream as he had not screamed since the truck ride from the crash site to the cellar. When he begged for morphine, Dr. Géri told him, "Is that the way the babies scream when you bomb them? Morphine is for heroes. Not for American fliers who bomb babies." Then he'd turn a wire tighter, and my father would scream louder. Dr. Géri was a pacifist. So strange, thought my father, to be tortured by the Allies, not the enemy. He had to love and hate Dr. Géri.

The Eighth Air Force continued to stage its bombing raids on the area, and when the bombers rumbled overhead, the light globe above my father's bed would start swinging as the 500-pounders detonated around the camp. He'd watch the light bulb and listen to his piano wires play a bizarre and dissonant tune, like Bartók; a prelude, it seemed, to a direct hit that would blow them all to bits. If the bombs were close enough, the light would swing so hard that it would shatter against the ceiling and shower him with broken glass.

By the time spring came to the German farm fields, my father's bones had knitted, and one of the very muscular male nurses, a French prisoner named Henri Moreou, would carry him upstairs like a baby and set him in the sun beneath a blanket. One day in April, he was sitting in his chair in the sun, the blanket over his knees. It was a perfect day, with just a few high clouds. A front had come through and cleared away the smells of war, which sometimes hung over them. The sun was warm and the air was cool. My father was left alone with the guards, who were scattered about some distance away. He watched the far hills, daydreaming and almost dozing off. Most of the dreams were about food. The German guards ate potatoes, which was all that was left in the war-torn countryside, and they gave the potato peelings to the prisoners to make a thin soup. Slowly starving, my father found that he had become obsessed with mayonnaise, which he loves to this day. At other times, he'd daydream about his mother, Rosa, who grew roses and painted and made pottery, or the girl back home, his fiancée and eventually my mother, Anna Marie Mosher (whose grandfather was a railroad worker and had been run over by his own train—on his sixtieth birthday).

My father would remember his old dog, whose name was GI, and his father, Agustín, coming home to Rosa after work, where he made barbecue in a stone pit over a mesquite wood fire and sold it to the workers in the area. Agustín would sweep the front porch and steps in the afternoon light and then continue sweeping down the sidewalk to the dirt street in the barrio where they lived, sweeping and sweeping, between the rows of Rosa's roses.

My father could hear the soft snap of playing cards to his left, where two guards were engaged in a game. To his right, some others were just standing, staring into space, and two more were sharing a cigarette. The hills were turning green. He felt calm, almost happy, and quite distant from the incessant pain of an empty stomach and knitting bones.

Something caught his eye on the top of the farthest hill. He saw something move. As he watched, a figure seemed to grow out of the hill. A man under a burden, walking, coming from the far side, now cresting the hill, now advancing over its near side. The figure was still too far away for my father to tell anything about it, but even at that distance, something about it struck my father as odd. Nothing ever came over those hills. And there was just something familiar in the movement. Impossible. He was too far away to distinguish any details except that the man labored under a large pack and other gear.

But my father was idle, dozing, and he had nothing to do other than watch as the figure came on and on. He didn't know how long he watched the figure grow out of the new green landscape. It fell into a depression between two hills and vanished for a while. Then it reappeared over the next rise, larger, more distinct, and my father knew that there was definitely something about it. He sat forward in his chair: something about the man's burden that my father just couldn't put his finger on. He wondered if he was hallucinating from starvation.

Then the guards noticed, too; the card game stopped and the others stood at the ready. A soldier ground out a cigarette with the toe of his boot and blew a thin stream of blue smoke into the windless air. His hand came up to shade his eyes as he watched. They formed a still tableau as the lone figure advanced across the hills, coming now through an open field of yellow flowers perhaps an eighth of a mile wide. He was dressed in green-gray, that much was now clear, and he was armed. The top of his head was round, and suddenly my father could see why: He wore a helmet. The field of yellow blossoms seemed so enormous and bright, as if the figure floated on a bowl of liquid sun.

The guards drew together into a group and placed their Schmeizers at the ready. Everyone was fixed so intently on that lone figure, that personage, arriving, arriving, taking so long to arrive, and the accumulation of detail and meaning as he grew larger, and the vast landscape around him and the yellow field of flowers that seemed alternately to swallow and offer him up as if he floated on an ocean wave. He might be an ant for all his mass, and yet how he commanded their attention, as if they were the members of a primitive cult awaiting at long last the returning god of their mythology.

Perhaps those boys knew long before my father did what they were looking at. The man was only 200 yards off when the injured flier began to put together in his mind what he was seeing. And yet his muddled mind would not believe it, and so he just stared dumbly.

Then he could hear the clanking, canteen and bayonet, the tinkling of dog tags, P-38, tin cup. All the gear made a sort of rattling, atavistic music, and the big rucksack shifted, boots shuffling, with that inimitable slack-limbed indolence—no, no one else can walk like that, pose like that; it was unmistakable, for there was only one breed of human being the whole world over who could be so unstrung yet graceful. Movies have been made, novels written, about nothing more than that insouciant walk, that very carefree nonchalance with which he ambled toward them, that cool.

He was a mere 50 yards off when my father's mind finally engaged, and at that moment he looked around at the guards, fully expecting to see them draw back the slides on their weapons and open fire. But what he saw instead was the young faces, upturned, the slack expressions, not of fear, but the relief of thank-God-it's-over, and as the lone figure advanced, they threw their weapons to the ground and put their hands in the air.

The single GI sauntered straight and cool and casual toward them, and the guards stood stock-still in their surrender, as the American Army scout crossed the compound yard toward my father, came right up to him, and cast his shadow over him so that my father could at last see his face—big, crooked teeth in a leather grin; soft, indefinite-colored hair falling across his tanned face under his helmet; M-1 rifle slung casually over his arm.

Chewing gum.

He grinned down, hardly glancing at the Germans. He shook a smoke out of a pack of Luckies and offered the wounded flier one. My father reached out and took it with his left, his good hand, and the GI flicked a Zippo with that inexpressible dexterity of the combat veteran and lit the two smokes, his own, then my father's, cupping his hands tenderly around my father's thin fingers. They both blew smoke out and stared at each other.

"Hello, GI," the GI said.

"You're a sight for sore eyes," my father said.

"You look like you could use a bite."

"Sure could." He was shaking all over, beset by a fever of unknown origin.

The GI dropped his pack, dug around in it, and came up with cheese and a chunk of coarse bread. He tore off some bread, handed it to the starving flier, and flicked out his gravity knife like a switchblade to cut a thick slice of cheese. My father fell to eating it like a dog, gnawing furiously, groaning out loud because he couldn't help himself, glancing up from bite to bite as if someone might snatch it from him. He noticed how sad the GI's grin was, and in it he saw how bad he must look, a ghost of himself, this flier in a threadbare uniform, torn and bloodstained. He had been taken prisoner weighing 170 pounds and went home at 119.

FAMILIES, TOO, develop their own survival rituals, their codes of integrity, ideas of what it means to be worthy. When I was growing up, my mother would make a special dinner every January twenty-third to celebrate the date my father was shot down, and each year I'd hear a little bit more of his improbable story. (I'd sometimes hear my father wake up screaming at night, too.) And although I knew the stories were true, I'm not sure I could ever quite square the image I had of that boy falling out of the sky with the man he would become. The proof was always before me. His right arm, the one that was fixed with a stainless-steel pin, moved only a few degrees at the elbow. When he dove off the diving board into the swimming pool, I could see how crooked it was. (Amazing that he could walk, let alone dive.) And when I was very little and came up only to his knees, I saw the horrible scars running the length of his shins. His feet were so deformed from the impact that they caused one of my brothers, Philip, to burst into tears as a toddler. My father had to have special shoes made just to walk without pain.

The lesson, which it took me many decades to learn, was that he was here among us because he was cool. He was cool now and had been cool at the moment of his death, saying nothing more than had to be said: "This is it," and, "Bailout, bailout, bailout," as prescribed on the checklist before him. As prescribed by the pilot's unspoken Stoic code of conduct. He received the Distinguished Flying Cross not for that last flight but for an earlier time when he was shot down and saved his crew through cool and skill and naked nerve. With two engines out, his radios gone, his plane's wings and tail shot to pieces, leaking fuel at a prodigious rate, he was inexorably descending through an overcast, recognizing that he'd have to order his crew to bail out, probably into the icy English Channel. They had no idea where they were, when he spied a rocket punching through the overcast and turned toward the pink glow. His wheels barked onto the asphalt runway somewhere in Belgium, just in time for everything on his airplane to quit. His happy crew partied there until dawn, when the sound of wooden clogs stirred them to head for bed as the local people went to work.

I have a photograph of him with three members of his crew, taken at the base in Nuthampstead before a flight in 1944. Charles Kahouri, who at that time was pilot to my father's co-pilot, stands on his right. To his left are Jack Layden and Jack Kutchback, both of whom flew the last mission. Those three men are neat and severe in their regulation uniforms, their hats on straight, their postures military. They look, well, nervous, if not afraid, even as they try to smile. My father, by contrast, is not only out of uniform, he has no shirt on. He wears Ray-Ban Aviators, his hat cocked at a rakish angle, one foot swung out before him as if he's about to do a little dance step. He's grinning like the devil that I'm told he was. I always looked at that photo and thought: What in hell was he thinking? Many years later, I looked at it again and realized that the other three were dead and he was alive.

He hadn't let his injuries stop him, either. Sunday morning, early, he'd suddenly appear in the kitchen with a top hat and cane, doing a soft shoe and singing, "Gimme that old . . . soft . . . shoe . . . ," making drum sounds and whistling the backup band arrangement. We'd squeal and clap, and then he'd twirl the cane around his finger like Diamond Jim the Riverboat Gambler. He'd throw down the cane, grab up three eggs from among the dozen my mother was about to cook for breakfast, and he'd begin juggling, even as she protested that if he broke them, he'd have to go out and get some more, and she wasn't about to clean this floor again, either.

Break them? Unthinkable.

Just to prove it, he'd juggle them behind his back. I had no doubt that he had been granted all of those abilities in one fell swoop by flying an airplane and being shot down. He had gone out to meet something terrible, and he had mastered it and had come back to be treated like a king by all those around him, to sit and smoke and to be suave, smart, handsome. The same innate focus and attention that kept him from dropping the eggs, that same ability to be an elite performer, had also allowed him to read the *Journal of Cell Biology* while five (and then six, and then seven) sons raged around him, wreaking havoc. That couldn't be any harder than reading an emergency checklist inverted at 27,000 feet with your left wing shot off while you were spinning hard enough to suck your eyeballs out.

I knew that there was little hope that I would ever have such righteous stuff. Certainly, he was never going to explain it to me. Aviators didn't chat like that. But the whole thing was irresistible. I was a child, but before I could even put a name on it I was determined to steal my share.

So it was that I ended up riding dirt bikes at 125 miles an hour on a dry lakebed in the Mexican desert in a whiteout dust storm. So it was that I wound up on a knife-edge cliff in a blizzard with no tent in the middle of the night on the highest point east of the Rockies. So it was that I found myself on a naked heap of chert somewhere above the Arctic Circle, clutching an automatic shotgun jammed with nine rounds of alternating double-ought buck and deer slugs, awaiting the approach of a grizzly bear who'd caught the scent of our fresh caribou meat. So it was that I wound up flying upside down, 10 feet off the ground, going 150 miles an hour, through an obstacle course in the Santa Susana Mountains in California. Then I'd write about it as best I could and give it to my father. Every ex-combat pilot has what they call an "I-Love-Me Room." In my father's den are his wings and memorabilia and the photos of him and his dead crew from the bad old Army Air Corps days. Across from that wall of glory, on a bookshelf, he keeps all the things I've written. My daughters tell me that I have the job every thirteen-year-old boy wants. My ex-wives tell me that I never grew up.

Once he was shot down, my father's survival was not a matter of crawling up a mountain or catching fish in the Atlantic, as it was for Joe Simpson or Steve Callahan. But I have to think that his whole life had led him to that one point in an unconscious sequence of circumstances, judgments, and acts, which combined in the thrall of the forces that Clausewitz called friction and chance, the bipolar pull that circumscribes and defines the universe. The road that leads a Japanese tourist to drive beneath a falling 30-ton rock in Glacier National Park stretches back to the first divisions of a zygote, even as it begins scrawling out the definition of itself in lines of sugarcoated DNA.

That doesn't mean everything is fated; indeed, just the opposite. It means the systems we live with are unpredictable and therefore have profound and unexpected results. But there are patterns in there, too. The same boy who rode his bicycle off a garage roof to see what would happen, who joined the cavalry in high school to feel the heat of the horse and the kick of the gun, had at last achieved what Leschak calls "an almost mystical plane of awareness" in learning to lean on the wind, accept the speed and noise and smoke, and to aim carefully and shoot straight while both calming and thrilling to the complex ballet of which he was the silent center, the jockey to the horse. To fly, then, he had to do the same again in the smell of oil, in the heat and smoke, and then once more teach his spirit to fly straight and level and calmly terrified while explosions rocked his ship and razor-sharp, red-hot fragments of supersonic flak penetrated the thin aluminum skin of his aircraft, punching smoky fingers of light into the darkness within. And when one of those fingers pointed out a man, it would mean to select him for sacrifice. The sweet, sharp, continuous anguish of such learning had allowed him to will himself alive in the impossible dream of air. "He worked out his own salvation."

Survival is a continuous spiritual and physical act that spans a lifetime. Riding his bicycle off the roof and all the rich spinning of a whirlwind childhood taught my father how to fall. Saving his crewmen in Holland made him worthy to lose them over Neuss. With good-hearted determination, he not only rebuilt his own life, he rebuilt his crew, siring eight sons. Sadly, the first died in infancy. But with my father as captain, our family made nine, which was the very number of men he had lost.

FIRST LIEUTENANT Federico Gonzales was liberated from the Gerresheim camp on April 17, 1945. It was almost thirty-four years later that I was writing for *Playboy* magazine, doing research on airline crashes and studying the flaws of one particularly notorious airplane, the McDonnell Douglas DC-10, a popular so-called jumbo jet that had suffered more catastrophic in-flight failures than any other modern jetliner. As a contributing editor for the magazine, I was planning to join my colleagues on a trip to the

American Booksellers Association Convention in L.A. Shel Wax, our managing editor, was going. His wife, Judy, was going with him to promote her first book, which had just been published. Our fiction editor, Vickie Chen Haider, was going, as well as our foreign rights editor, Mary Sheridan. I was planning to join them on American Flight 191 to Los Angeles on the afternoon of May 25, 1979. But when I found out the airplane was a DC-10, I told Shel I'd thought better of it. He laughed and said I'd been reading too much. He was right, I had. Although I'd been flying in and out of crowded airspace in a small Piper aircraft for several years by then, the idea of getting on a DC-10 terrified me.

That morning, I sat in Shel's office on the tenth floor of the old Palmolive Building, where *Playboy* had its headquarters. I was talking to Judy, who was a good friend. She signed a copy of her book for me. I said good-bye to Vickie, who had a one-year-old son. She and I often rode the bus to work together. I stopped in to see Mary, too, and wish her a good trip. I watched Shel and Judy go out to the Art Deco elevators walking arm in arm. I remember thinking how cool it was that they were still so in love, whispering and laughing like teenagers as they waited for the elevator.

The flight lasted thirty-one seconds and crashed in an open field, just missing a fuel-tank farm and a trailer park. The plane rolled nearly inverted before it hit the ground. Everyone was killed, 273 people, making it the worst aviation disaster in American history even now, nearly a quarter century later. I lived only twenty minutes from the crash site and was there to report on it just after the fire was put out. Vickie, beautiful Vickie, with her straight black Chinese hair, had to be identified by a bit of dental-work.

The event launched me into an even more intense period of flying and writing about aviation. But I was always haunted by how close I'd come to making my life exactly match my father's. I had always followed him, followed his example, tried to be like him. I thought of myself as the hero's apprentice. But later on, I began to see that I had it all wrong. He was no hero. He was a survivor. And somehow I had worked out my own salvation, my survival, in a long series of acts, conditions, and judgments leading up to the single word I spoke to Shel when he found me sitting on his raw silk couch with his wife and asked me if I didn't really want to come with them to L.A. that afternoon. My answer was: No. I had come to be a survivor, too, and not even the old man was the old man any more.

ALL OF the acts, conditions, and judgments of a lifetime had put my father on a vector toward a spot in space and time where an 88-millimeter shell happened to be rising toward 27,000 feet above mean sea level on January 23, 1945. People have long accepted, at an unconscious level, the essence of theories such as chaos and complexity. Many stories have been written about what would happen if you could travel back in time and change just one thing, no matter how trivial. The doggerel verse that begins: "For lack of a nail a shoe was lost/ For lack of a shoe a horse was lost . . ." captures the idea. If Colonel Hunter had elected to fly left seat instead of right that day, I would not have been born, and you would not be reading this book. If I had been assigned to another story in 1973 instead of airline safety, I wouldn't have known about the DC-10 and would have gotten on that plane with Shel and Judy. And you would not be reading this book.

But survival in the moment, or over hours or days or months, whether that survival comes about by chance or effort or an inex- plicable combination, must be followed once more by the same struggle that led to that point. As Solon pointed out to Croesus, a life cannot be judged until it is complete. My own survival in not going with Shel and Judy, Vickie and Mary, and in all sorts of other situations, is something I'm still working out. If my father's fall planted the seeds of this book, then the crash of American Flight 191 fertilized them and made them grow. In a world governed by an ineluctable order, which pushes through Newtonian physics, Einsteinian relativity, thermodynamics, and quantum theory with all the certainty of gravity or any other encroaching natural law, nothing can truly be said to happen by chance, which is just a word we invented to explain the troublesome boundary between order and chaos. Fate, then, turns out to be the struggle, the tension, between the natural law that dictates that everything should proceed toward disorder (entropy) and the natural law that dictates that everything should be self-organizing (complexity theory). If those are, indeed, the two overarching natural laws, then everything becomes clear and we go forward into the past to find the Chinese concept of yin and yang.

Certainly, my father's survival did not end with his falling from the sky. I watched it take shape, even as it shaped me and my world. It began there, a man with broken legs and broken arms and broken feet and ribs, his nose stuck back on almost as an afterthought by a boy who happened by as he was weeping. Then he was packaged and shipped home. (He told me that the most fearsome flight he'd ever had was not when his wing was shot off. It was the flight home when they encountered a thunderstorm and he sat watching the wings make wild excursions up and down, emptying the ashtrays on that old DC-3.)

He picked himself up and strove endlessly to grasp the world in which he found himself. I saw him rise from the grave and earn a Ph.D., find a job at a prestigious medical school, publish scientific papers, send platoons of new doctors out the door to heal, and in his spare time, learn to become an excellent potter, to paint and draw and sing and play piano, carve sculptures out of wood, build model planes, tinker together our first stereo set, and drive his noisy family all over the continent in a 1956 Volkswagen bus looking for adventure. I saw him constantly and hungrily grappling with his world, trying everything, sampling everything, tasting the world, to understand, to feed his insatiable curiosity, even as he sat in darkness and peered through an electron microscope at the inner secrets of a cell.

We spent one whole summer carving boomerangs out of various kinds of wood and studying the aerodynamics to explain why they returned instead of doing what Newton said they'd do: keep going.

He was the only man I knew who'd read *Finnegans Wake* from cover to cover. He reminded me of the Great Santini, who told his son, "Eat Life, or Life will eat you." In his Zen fashion, my father would say, when I did something inexplicably wild, "Okay, but if you break your leg, don't come running to me."

I saw that catastrophe had not broken him. He was the student who learned how to duck and therefore no longer needed swordsmanship. Adversity annealed him. It gave him endless energy. He taught me the first rule of survival: to believe that anything is possible.

THE RULES OF ADVENTURE

AT THIS POINT you may be looking back over this book and thinking: That's all well and good, but what do I do now? How do I avoid getting into a survival situation; and if something should happen, how do I get out again? This book is not meant to tell people what to do but rather to be a search for a deeper understanding that will allow them to know what to do when the time comes— and it always comes, in some form, for all of us.

Nevertheless, there is some advice that can help in any survival situation, whether it's emotional survival, financial survival, or survival after breaking your leg at 19,000 feet on a Peruvian mountain. The concepts can be applied to any stressful situation, not just ones in which your life is threatened. These concepts, like this entire book, are not necessarily meant for the elite mountaineer, the fighter pilot, or the Olympic kayaker, who face death on their own terms. They are meant for the rest of us, who just want to get through life's rough spots and go out now and then to see the beauty or to get a little adrenaline fix.

One piece of advice is to take the training that's available.

Although most survival schools may not teach the mysteries of survival or accident theory, they have a lot to offer, and I recommend them, even to people who aren't planning to go into the wilderness. Once I was at Massacre Lake in northwestern Nevada, and a Bureau of Land Management geologist told me about a young couple from California with a new baby. They were traveling east and made the mistake of driving up a back road and into a blizzard in the mountains. They were stranded in a cave for several days until the man was able to walk out and get help. They weren't planning to go into the wilderness, but when the thin reed of technology on which they were leaning so heavily failed them, that's where they found themselves. Whatever else they did right or wrong, they would have done better if they'd been to a couple of survival schools first. At least they would have known to carry firemaking materials.

I believe everyone should learn about basic survival skills and the survivor's frame of mind, because they come in handy when the trappings of civilization (or even the financial or emotional support) that we take for granted drop away for whatever reason. Instructors at survival schools are finding that more and more business people are taking their courses, not because they want to survive in the wilderness but for the other qualities that grow out of knowing how. Here are a few suggestions, first for staying out of trouble, then for dealing with it when it comes.

Perceive, believe, then act. Avoiding accidents, avoiding survival situations, is all about being smart. Horace Barlow, a neurobiologist, says that intelligence is a matter of "guessing well." Guessing well involves a natural tendency people have: to predict. Training is an attempt to make predictions more accurate in a given environment. But as the environment changes (and it always does), what you need is versatility, the ability to perceive what's really happening and adapt to it. So the training and prediction may not always be your best friend.

The Korean martial art called Kum Do involves very sharp swords and (at least originally) an eventual fight to the death. It is related to Japanese Kendo. Those who practice it today use bamboo sticks, but some of the moves that have survived the centuries were developed to shake blood off the blade so that it wouldn't coagulate and dull the edge. Kum Do teaches students to avoid the "Four Poisons of the Mind"—fear, confusion, hesitation, and surprise. I had to think about that last one for a while, as the others seemed self-evident to me. But I recognized that our constant tendency to anticipate and predict may sometimes put us at a disadvantage. In Kum Do, the student must not anticipate his opponent's moves or allow his natural instinct for prediction to run free, for that could lead to surprise, which could lead to momentary confusion and then sudden death. Instead, he must watch, clear and calm, and then act decisively at the correct moment. Since a single blow can be lethal, can be in fact the only blow dealt in the fight, Kum Do teaches focus, precise bodily control, courtesy, humility, and confidence. Those are similar to the qualities you need when you engage the forces of nature.

Some instructors at survival schools use the acronym STOP—stop, think, observe, plan. (It ought to be stop, observe, think, plan, and act, but that wouldn't make such a cute acronym.) Those instructors intend to teach you how to react during a survival situation, but the same skills can help you to avoid one.

It's important to have a plan and a backup plan or a bailout plan. What-if sessions can help to develop backup plans and should precede any hazardous activity. But you must hold onto the plan with a gentle grip *and* be willing to let it go. Rigid people are dangerous people. If you plan all your moves beforehand in Kum Do, you will be cut in half by your opponent's sword. Survival is adaptation, and adaptation is change, but it is change based on a true reading of the environment. Sometimes sticking to the plan is best and improvising causes trouble. When I went into Glacier National Park (chapter 9), if I had simply stuck to the plan to walk the nature trail instead of getting sucked into a long hike by the beauty, I would have stayed out of trouble. And if I had turned around each time I reached a fork in the path and carefully observed what the correct trail looked like, I would not have gotten lost.

Those who avoid accidents are those who see the world clearly, see it changing, and change their behavior accordingly. This won't save everyone from everything. Nothing will. But it will help a great deal in most situations. As Mark Morey, my survival instructor from Vermont, told me, "We come from cities and learn to expect things to stay the same. But they don't. And it kills us, quickly or slowly."

Avoid impulsive behavior; don't hurry. Catecholamines are a double-edged sword. They provide power when you need that burst of energy. (I know a man who lifted a full-size Chrysler four-door sedan off his daughter's leg after a car accident.) But they can also make you so excited that you do the wrong thing. Don't be the snowmobiler or skier who dashes out onto an avalanche slope just because it's a wonderful day and the beauty is on duty.

John Bouchard is an elite mountaineer with more than thirty years of experience. He told me, "On my last expedition in Pakistan, it was obvious that conditions weren't right. We'd tried [the climb], and our ropes were cut by stonefall. I was shaken up, and I had a bad feeling. I didn't want to go. But my partner insisted. So we tried another route, and once I started, it was like a dog after a bitch in heat. I forgot my resolve to be careful. I just said, 'Screw it, let's go for it.' We didn't even take any bivy gear. The strange thing is, I know I'd do it again." Here he laughed, still unable to explain to himself how someone with so much knowledge and experience could have done something so foolhardy. "Luckily, in the last four hundred feet we didn't have the right gear. There was no way of protecting the climb, so we had to turn around at twenty thousand feet. We didn't have to spend the night out there."

Even elite performers struggle to find emotional balance and control. They, too, have to correct their courses as they go. They make big mistakes, or small ones with big consequences; what separates the living from the dead is an ability to see the error and adapt, a determination to get back on the path. Even the most unforgiving environments allow a few sins if you adjust your behavior and take correct action in a timely fashion.

Know your stuff. Marcus Aurelius, the Roman emperor and general, wrote: "Of each particular thing, ask: 'What is it in itself, in its own construction?'" A deep knowledge of the world around you may save your life. People who don't understand the nature of water, for example, may tie a rope around themselves, as they do too frequently before entering swift water—to save someone, to retrieve something, to cross to the other bank. And then they die, because the system created by water and rope pulls them under and holds them there. The climbers on Mount Hood did not understand the system they were using nor the forces they could encounter (chapter 6). Know the system and keep in mind that the forces may be so large (or fast) that they're difficult to imagine.

Get the information. When I met Mike Crowder (chapter 8) on Lydgate Beach on Kaua'i, I was dangerously ignorant. Mine was not an emotional problem. I wasn't impulsively going to leap into the surf there. I just wanted to go swimming and didn't have the information. There are lots of places like that—Mount Hood, Mount Washington, Lambs Slide, Glacier National Park, the Potomac River, and most ski areas. Those are what I call danger, or hazard, zones. The same accidents happen over and over, year after year. It's a simple thing to know, but so many people plunge in without inquiring. Park rangers, lifeguards, and local authorities are happy to tell you if you only ask.

Commune with the dead. If you could collect the dead around you and sit by the campfire and listen to their tales, you might find yourself in the best survival school of all. Since you can't, read the accident reports in your chosen field of recreation. *Accidents in North American Mountaineering*; the National Speleological Society's newsletter; *River Safety Report*; and numerous other publications (such as the *Morbidity and Mortality Weekly Report*) and Web sites not only provide reading that is by turns gripping, hilarious, and heart-wrenching, but also tell you the mistakes other people have made. Then you can be on the lookout for similar situations and perhaps avoid them. (Wasn't this the cave where those scuba divers drowned . . . ?)

Be humble. A Navy Seal commander told Al Siebert, the psychologist who studies survival, that "the Rambo types are the first to go." Don't think that just because you're good at one thing, it makes you good at other things. People in business make that mistake all the time, as the history of Xerox, A&P, Digital Equipment Corporation, and numerous other companies demonstrate. Captain Gabba (chapter 3) was no doubt a fine Army Ranger. But as much as he knew, he didn't know enough about fast-moving rivers (and he didn't know that he didn't know), so he drowned.

The other side of this principle is the danger of success. John Bouchard, who is also a paraglider pilot, describes it: "You have one hundred hours. You think you understand it. You know how to thermal, you've done some cross-country trips, you've been up to the cloud base, you've flown twenty miles, and you think you understand. Then you push it. Look out!"

During the 1992 National Paragliding Championships at Owens Valley in California, his friend, Leo, had his first big flight two days before practice, with an 8,000-foot gain and 40 miles cross-country from Bishop to Mina, Nevada. "The next day," Bouchard said, "he was low, trying to catch a thermal, and it was a big thermal. It tipped the glider, and he spun out of control. He tried to recover rather than throw his reserve chute, and he hit the ground and died. If he had not had that unbelievable flight the day before, I think he would have thrown his reserve." Leo had developed a new secondary emotion and it killed him. "These sports are different in that you need a mentor: You need it more after you've learned the basics than at first. In climbing, you find a mentor and attach yourself to him. Someone who's been there. Part of knowing yourself is through your mentor."

Those who gain experience while retaining firm hold on a beginner's state of mind become long-term survivors. In flying, we constantly go back and retrain to remind ourselves that Newton's laws don't get suspended just because we've become so goddamned studly. That's hard to remember when you feel like all you have to do is find a phone booth, get that cape on, and leap out the nearest window. We have a saying: "There are old pilots and bold pilots, but there are no old, bold pilots." That's true in all hazardous pursuits.

When in doubt, bail out. This is a tough one. You've paid for airfare. You've waited all year for this trip. You've bought all your equipment. It's hard to admit that things aren't going to go your way. At times like that, it's good to ask yourself if it's worth dying for. Of course, some people need to get close to death, to take the really big risk. (Or as a friend of mine who is a Chicago cop once told me, "Some people just want to die in a hail of police gunfire.") But as Boone Bracket used to say, "I'd much rather be on the ground wishing I were in the air than in the air wishing I were on the ground." Gary Hough (chapter 4) pulled out of the Illinois River when he saw 18-inch trees shooting past him at 15 miles an hour, while others drowned. David Stone (chapter 5) was struck by lightning on the southeast buttress of Cathedral Peak because of an inflexible plan, a failure to have a bailout plan, and perhaps even a bit of difficulty with impulse control, a difficulty that all of us who pursue adventure have.

Drew Leeman is director of risk management for the National Outdoor Leadership School, which trains the guides and guides the trainers in anything from whitewater kayaking to snowboarding. "You need humility," Leeman said, in order to resign in time. "We say at NOLS, 'The summit is not the only place on the mountain.' Maybe you've tried to climb this peak for five years and maybe you're on the third try and the weather is getting lousy again."

Some people get caught up in thinking, "I just spent all my money to get here and I'm not going to let some storm ruin it for me," Leeman said. "At that moment, you need to be able to look within yourself and say, 'It's not going to happen this year, either.' Otherwise, you run the risk of going for it and having unfortunate consequences. It's a matter of looking at yourself and assessing your own abilities and where you are mentally, and then realizing that it's better to turn back and get a chance to do it again than to go for it and not come back at all." We are a society of high achievers, but in the wilderness, such motivation can be deadly sometimes. "Be realistic about your goals and your time frame," Leeman advised. "Then be content with being outdoors. If you can get to the summit, that's just the icing."

WHEN TROUBLE comes, in whatever form, the same rules apply and a few others, too. You have to have that Positive Mental Attitude. Blaming others is a national pastime with the frivolous lawsuits to prove it. That's not a survivor's attitude, but Steve Callahan's observation that he had a view of heaven from a seat in hell was. He was looking at the glass as half full.

Regarding self pity, Stockdale reflected on the value of having read the Book of Job: "When the Lord appeared in the whirlwind, he said 'Now, Job, you have to shape up! Life is not fair . . . ' This was a great comfort to me in prison. It answered the question 'Why me?' It cast aside any thoughts of being punished for past actions." But that attitude won't magically appear when you need it. It has to develop throughout life as a way of looking at the world, and it's worth working on, since it sets you up for success in any situation. It sets you up to be a rescuer, not a victim.

Jim Stockdale believed, as I do, that the Stoic philosophers had it right when it came to survival as a way of life. How else could you describe Joe Simpson and Steve Callahan's attitude if not Stoic and philosophical? Stockdale also recommends "a course of familiarization with pain." Some survival schools can provide that. Trips into the wilderness can provide it unexpectedly, too. By embracing the natural pain that occurs in everyday life, you are practicing for a survival situation. A survival situation may occur in the wilderness, but it may also occur as war, illness, business failure, the need to get out of a burning building, and in numerous other forms. The reason Marcus Aurelius sounds as if he's writing a survival manual is because, at the time he wrote his *Meditations*, he was the general in charge of a very trying war, which he ultimately won.

In summarizing what Marcus Aurelius had to say, Irwin Edman, in his introduction to the *Meditations*, wrote: "Fortitude is necessary, and patience and courtesy and modesty and decorum, and a will, in what may for the *moment* seem to be the worst of worlds, to *do* one's best" (his emphasis). He wrote those words during World War II, when the Allies were in a desperate fight for survival. The same applies to any survival situation.

I'VE BEEN reading accident reports of various kinds for thirty or more years. Call me callous, but to me they're like silent comedy movies. People do the strangest things and get themselves into the most amazing predicaments. You want to go wake up Tolstoy and Dostoevsky and say: Hey, you think your characters are crazy. . . .

In reading about cases in which people survived seemingly impossible circumstances, however, I found an eerie uniformity. Decades and sometimes even centuries apart, separated by culture, geography, race, language, and tradition, they all went through the same patterns of thought and behavior. I eventually distilled those observations down to twelve points that seemed to stand out concerning how survivors think and behave in the clutch of mortal danger. Some are the same as the steps for staying out of trouble. Here's what survivors do:

1. Perceive, believe (look, see, believe). Even in the initial crisis, survivors' perceptions and cognitive functions keep working. They notice the details and may even find some humorous or beautiful. If there is any denial, it is counterbalanced by a solid belief in the clear evidence of their senses. They immediately begin to recognize, acknowledge, and even accept the reality of their situation. "I've broken my leg, that's it. I'm dead," as Joe Simpson (chapter 13) put it. They may initially blame forces outside themselves, too; but very quickly they dismiss that tactic and recognize that everything, good and bad, emanates from within. They see opportunity, even good, in their situation. They move through denial, anger, bargaining, depression, and acceptance very rapidly. They "go inside." Bear in mind, though, that many people, such as Debbie Kiley (chapter 11), may have to struggle for a time before they get there.

2. Stay calm (use humor, use fear to focus). In the initial crisis, survivors are making use of fear, not being ruled by it. Their fear often feels like and turns into anger, and that motivates them and makes them sharper. They understand at a deep level about being cool and are ever on guard against the mutiny of too much emotion. They keep their sense of humor and therefore keep calm.

3. Think/analyze/plan (get organized; set up small, manageable tasks). Survivors quickly organize, set up routines, and institute discipline. In successful group survival situations, a leader emerges often from the least likely candidate. They push away thoughts that their situation is hopeless. A rational voice emerges and is often actually heard, which takes control of the situation. Survivors perceive that experience as being split into two people and they "obey" the rational one. It begins with the paradox of seeing reality—how hopeless it would seem to an outside observer—but acting with the expectation of success.

4. Take correct, decisive action (be bold and cautious while carrying out tasks). Survivors are able to transform thought into action. They are willing to take risks to save themselves and others. They are able to break down very large jobs into small, manageable tasks. They set attainable goals and develop short-term plans to reach them. They are meticulous about doing those tasks well. They deal with what is within their power from moment to moment, hour to hour, day to day. They leave the rest behind.

5. Celebrate your successes (take joy in completing tasks). Survivors take great joy from even their smallest successes. That is an important step in creating an ongoing feeling of motivation and preventing the descent into hopelessness. It also provides relief from the unspeakable stress of a true survival situation.

6. Count your blessings (be grateful—you're alive). This is how survivors become rescuers instead of victims. There is always someone else they are helping more than themselves, even if that someone is not present. One survivor I spoke to, Yossi Ghinsberg, who was lost for weeks in the Bolivian jungle, hallucinated about a beautiful companion with whom he slept each night as he traveled. Everything he did, he did for her.

7. Play (sing, play mind games, recite poetry, count anything, do mathematical problems in your head). Since the brain and its wiring appear to be the determining factor in survival, this is an argument for expanding and refining it. The more you have learned and experienced of art, music, poetry, literature, philosophy, mathematics, and so on, the more resources you will have to fall back on. Just as survivors use patterns and rhythm to move forward in the survival voyage, they use the deeper activities of the intellect to stimulate, calm, and entertain the mind. Counting becomes important, too, and reciting poetry or even a mantra can calm the frantic mind. Movement becomes dance. One survivor who had to walk a long way counted his steps, one hundred at a time, and dedicated each hundred to another person he cared about.

Stockdale cites "love of poetry" as an important quality for enduring. "You thirst to remember," he wrote. "The clutter of all the trivia evaporates from your consciousness and with care you can make deep excursions into past recollections. . . . Verses were hoarded and gone over each day. . . . [T]he person who came into this experiment with reams of already memorized poetry was the bearer of great gifts."

Survivors often cling to talismans. They search for meaning, and the more you know already, the deeper the meaning. They engage the crisis almost as a game. They discover the flow of the expert performer, in whom emotion and thought balance each other in producing action. "Careful, careful," they say. But they act joyfully and decisively. Playing also leads to invention, and invention may lead to a new technique, strategy, or a piece of equipment that could save you.

8. See the beauty (remember: it's a vision quest). Survivors are attuned to the wonder of the world. The appreciation of beauty, the feeling of awe, opens the senses. When you see something beautiful, your pupils actually dilate. This appreciation not only relieves stress and creates strong motivation, but it allows you to take in new information more effectively.

9. Believe that you will succeed (develop a deep conviction that you'll live). All of the practices just described lead to this point: Survivors consolidate their personalities and fix their determination. Survivors admonish themselves to make no more mistakes, to be very careful, and to do their very best. They become convinced that they will prevail if they do those things.

10. Surrender (let go of your fear of dying; "put away the pain"). Survivors manage pain well. Lauren Elder (chapter 13), who walked out of the Sierra Nevada after surviving a plane crash, wrote that she "stored away the information: My arm is broken." That sort of thinking is what John Leach calls "resignation without giving up. It is survival by surrender." Joe Simpson recognized that he would probably die. But it had ceased to bother him, and so he went ahead and crawled off the mountain anyway.

11. Do whatever is necessary (be determined; have the will and the skill). Survivors have meta-knowledge: They know their abilities and do not over- or underestimate them. They believe that anything is possible and act accordingly. Play leads to invention, which leads to trying something that might have seemed impossible. When the plane in which Lauren Elder was flying hit the top of a ridge above 12,000 feet, it would have seemed impossible that she could get off alive. She did it anyway, including having to down-climb vertical rock races with a broken arm. Survivors don't expect or even hope to be rescued. They are coldly rational about using the world, obtaining what they need, doing what they have to do.

12. Never give up (let nothing break your spirit). There is always one more thing that you can do. Survivors are not easily frustrated. They are not discouraged by setbacks. They accept that the environment (or the business climate or their health) is constantly changing. They pick themselves up and start the entire process over again, breaking it down into manageable bits. Survivors always have a clear reason for going on. They keep their spirits up by developing an alternate world made up of rich memories to which they can escape. They mine their memory for whatever will keep them occupied. They come to embrace the world in which they find themselves and see opportunity in adversity. In the aftermath, survivors learn from and are grateful for the experiences they've had.

JOHN GRAY is the only guide licensed to take backpackers into Glacier National Park. His company, Glacier Wilderness Guides, runs the Granite Park Chalet there. The Highline Trail is a very popular spot in the park. People set off from their Winnebagos in the vast Logan Pass Visitor Center parking lot, a place where it can snow 12 inches in August. They walk with their kids and their cameras right out along the Continental Divide for the beautiful views. "They're just clueless when they start," Gray told me. "They don't even realize that being in the mountains you have to be prepared. A ton of people take off up there without proper equipment, and it rapidly becomes a life-threatening situation. This year at Granite Park Chalet we gave away every garbage bag we had to people who had come up without proper clothes and were hypothermic. We could warm them up, but the garbage bag was the only thing we could give them for the walk back down."

"There's the allure of the unknown and that can be compelling," Drew Leeman said. "To just want to go off without knowing what's out there. But it's important to do some homework first and to understand what kind of conditions you might be going into. That will determine your equipment, your route, your goals, and the amount of time you allow to achieve them."

Gray, Leeman, and others emphasize the essential tools of adventure: planning, caution, training, and learning some good decision-making skills.

The astronauts provide us with an excellent example. Never has there been a more celebrated group of thrill-seekers. Yet they more than any other single group understand that their lives are not their own to throw away. We are a nation of adventurers, and they belong to us, not only because we pay their way but also because they are the embodiment of that deep, abiding spirit of adventure and exploration that has always been essential to our national identity. They engage in an activity that is predictably lethal: Some of them are going to die, and they're going to do it on a fairly regular schedule. But when they do, we can feel secure in celebrating their lives and the assiduous attention to detail that went into their efforts to come back safely.

Everyone is a hero to someone back home. If someone doesn't come back, then others, in the form of search and rescue crews, will have to put their lives at risk during a rescue or body recovery. The real heroism of the astronauts is not in the risks they take—any idiot can throw his life away—but in the much more arduous lengths to which they go to protect themselves from harm. They know that adventure becomes folly when you stop looking out for number one. It's not selfishness, it's that fine distinction between going forth boldly and going forth blindly, a balance between dedication to the mission and informed caution. It takes real skill to strike that balance, and at their best, adventurers are both bold and cautious. That means knowing where the envelope ends, and moreover, knowing yourself well enough to estimate correctly just how far beyond it you can go and still get back.

I've watched as astronauts such as Bob Cabana, Gerry Ross, and Jim Newman, who did the first assembly of the International Space Station, become masters of their mission. They studied everything, practiced every move, and trained half a decade for a task that took two weeks. It's the secret of their success. By the time they get to the real mission, it seems like just one more drill: It's familiar, and that allows clear thinking. A panicked mind is a useless mind.

The outcome of a survival situation depends largely on your mental, emotional, and physical condition and activities. Everyone who meets catastrophe or challenge and survives it through his or her own actions goes through an initial transformation from victim to survivor, and also follows a well-defined pattern of mental and emotional checks, controls, actions, and transformations. Those activities, such as the split of the rational from the emotional self and the sudden, almost blinding insight that one is going to live, are far more important in predicting survival than any particular skill, training, or equipment. Those mental processes and transformations reflect actual brain activity that scientists are just beginning to understand. People who engage in wilderness recreation or risky outdoor sports can benefit from learning about such processes and transformations. Everyone has finite resources going into a catastrophe. It is in managing those resources and taking advantage of every bit of luck that comes along that survivors have been able to bring out their stories. There is no telling how many people behave and adapt as perfect, textbook survivors, only to die owing to extreme objective hazards that even the best behavior cannot overcome. In other words, you may do everything right and still die. Likewise, you may do everything wrong and live, as so many do every day.

NOT LONG ago, when I was flying the aerobatics contest circuit, a friend of mine, Jan Jones, was going about her ordinary business. She had started with the International Aerobatics Club at about the same time I did. Jan was an otherwise fairly ordinary person: She went to the grocery store, took a shower every morning, watched TV, did laundry. She and her husband went to the movies. Like me, Jan was a judge, and sometimes we'd sit together while others flew. Then one day she took off, crashed her Pitts, and died. People talked gravely about the unnecessary risks she'd taken. But another friend of mine was doing the very same everyday errands and activities but not flying, not taking the risks. In fact, she lived an extremely quiet life. Then one day she felt ill and went to the doctor and a few weeks later was dead from glioblastoma. Survivors know, whether they are conscious of it or not, that to live at all is to fly upside down (640 people died in 1999 while choking on food; 320 drowned in the bath tub). You're already flying upside down. You might as well turn on the smoke and have some fun. Then when a different sort of challenge presents itself, you can face it with the same equanimity.

D. H. Lawrence wrote that every year you pass an anniversary unaware: the anniversary of your own death. I've seen it so many times before, as adventurers circle and circle the spot marked X where they meet their own death, taunting it, teasing it, playing with the big cat. Never fool yourself into thinking you can tame it.

Sure, we do it in our ordinary lives, too—you can get killed on your lunch break—but we do that blindly. In our adventures, we engage fate deliberately. We choose a relentless and indefatigable opponent, while others pretend to be safe. We feel that our experiences are much more real, while seeing the masses as deluded in their complacency. When well-trained people are fetched off by fate during a well-planned and thoughtful expedition, there is no more ignominy in it than when an ordinary Joe gets hit by a bus. No one says, "He shouldn't have been walking there." But a climber named Karl Iwen, unfamiliar with Three Fingered Jack, a volcanic mountain in Oregon, which he was descending, left his companions, left the trail, left his ice ax strapped to his pack, and ventured out onto the snow, where he treated his companions to a spectacular show as he slid into the couloir and did a 600-foot Peter Pan. Karl did not die doing what he loved. He died of poor impulse control, or what I call "the rapture of the shallow."

The perfect adventure shouldn't be that much more hazardous

in a real sense than ordinary life, for that invisible rope that holds us here can always break. We can live a life of bored caution and die of cancer. Better to take the adventure, minimize the risks, get the information, and then go forward in the knowledge that we've done everything we can.

No, some people would rather not see it, but the bull is there for all of us. Some of us choose to pass the cape in front of its horns. To live life is to risk it. And when you feel the rush of air and catch the stink of hot breath in your face, you enter the secret order of those who have seen their own death close up. It makes us live that much more intensely. So intense is it for some that it seals their fate; once they've tasted it, they just can't stop. And in their cases, perhaps we have to accept that the light that burns brightest burns half as long.

But I believe that if you do it right, you can have it all. I adhere to what my daughter Amelia calls the Gutter Theory of Life. It goes like this: You don't want to be lying in the gutter, having been run down by a bus, the last bit of your life ebbing away, and be thinking, "I should have taken that rafting trip . . ." or, "I should have learned to surf . . ." or "I should have flown upside down—with smoke!"

Pete Conrad was the third man to walk on the moon. He died in a motorcycle accident on an ordinary day. It took him a while to die as he went to the hospital. I wonder what he was thinking. I hope it was: *I did it all.*

SELECTED
BIBLIOGRAPHY

Anderson, Scott, et al. "War Comes to America," *Esquire* (November 2001).

Apter, Michael. *The Psychology of Excitement*. New York: Macmillan Publishing Company, 1992.

Ashcraft, Tami Oldham, with Susea McGearhart. *Red Sky in Mourning: A True Story of Love, Loss, and Survival at Sea*. New York: Hyperion, 2002.

Aurelius, Marcus. *Meditations,* trans. George Long and ed. Irwin Edman. New York: Walter J. Black, 1945.

Bailey, Maurice, and Maralyn. *117 Days Adrift*. Dobbs Ferry, NY: Sheridan House, 1992.

Barnhart, Robert K. *Dictionary of Etymology*. New York: HarperCollins, 1995.

Barzun, Jacques. *Science: The Glorious Entertainment*. New York: Harper & Row, 1964.

Bluhm, H. H. "How Did They Survive?" *American Journal of Psychotherapy*, vol. 2, no. 1 (1948).

Burney, Christopher. *Solitary Confinement*. New York: St. Martins, 1961.

Callahan, Steven. *Adrift: Seventy-six Days Lost at Sea.* New York: Ballantine Books, 1986.

Chaikan, Andrew. *A Man on the Moon: The Voyages of the Apollo Astronauts.* New York: Penguin Books, 1994.

Charrière, Henri. *Papillon,* trans. June P. Wilson and Walter B. Michaels. New York: Morrow, 1970.

Chatwin, Bruce. *The Songlines.* New York: Viking Penguin, 1988.

Cherry-Garrard, Apsley. *The Worst Journey in the World.* Santa Barbara, CA: Narrative Press, 1922.

Clausewitz on Strategy: Inspiration and Insight from a Master Strategist, trans. William Skinner and ed. Tiha von Ghyczy, Bolko von Oetinger, and Christopher Bassford. New York: John Wiley & Sons, 2001.

Clifford, Hal. *The Falling Season: Inside the Life and Death Drama of Aspen's Mountain Rescue Team.* Seattle, The Mountaineers, 1998.

Collins, Jim. *Good to Great.* New York: HarperCollins, 2001.

Conrad, Joseph. *Lord Jim.* New York: Oxford University Press, 1999.

Damasio, Antonio. *Descartes' Error: Emotion, Reason, and the Human Brain.* New York: HarperCollins, 1994.

————. *The Feeling of What Happens: Body and Emotion in the Making of Consciousness.* San Diego: Harcourt, 1999.

Dostoevsky, Fyodor. *Memoirs from the House of the Dead,* trans. Jessie Coulson and ed. Ronald Hingley. New York: Oxford University Press, 1991.

Drury, Bob. *The Rescue Season.* New York: Simon & Schuster, 2001.

Elder, Lauren, with Shirley Streshinsky. *And I Alone Survived.* New York: Thomas Congdon, 1978.

Epictetus. *Enchiridion,* trans. George Long. New York: Prometheus Books, 1991.

Faith, Nicholas. *The World the Railways Made.* New York: Carroll & Graf, 1990.

————, and Doug Fessler. *Snow Sense*. Anchorage, Alaska: Alaska Mountain Safety Center, 1999.

Fussell, Paul. *Wartime: Understanding and Behavior in the Second World War*. New York: Oxford University Press, 1989.

Ghinsberg, Yossi. *Back from Tuichi*. New York: Random House, 1993.

Gladwell, Malcolm. "The Art of Failure," *The New Yorker*, August 21 and 28, 2000.

————. "Wrong Turn," *The New Yorker*, June 11, 2001.

Gleick, James. *Chaos*. New York: Penguin Books, 1987.

Glynn, Ian. *An Anatomy of Thought: The Origin and Machinery of the Mind*. New York: Oxford University Press, 1999.

Goleman, Daniel. *Primal Leadership: Realizing the Power of Emotional Intelligence*. Cambridge, MA: Harvard Business School Press, 2002.

Gray, Mike. *Angle of Attack: Harrison Storms and the Race to the Moon*. New York: Norton, 1992.

Haines, R. F. "A Breakdown in Simultaneous Information Processing." In G. Obrecht and L. Stark, eds., *Presbyopia Research*. New York: Plenum, 1991.

Hampden-Turner, C. M. *Maps of the Mind*. New York: Macmillan Publishing Company, 1981.

Hardcastle, Nate, ed. *Survive: Stories of Castaways and Cannibals*. New York: Thunder's Mouth Press, 2001.

Herodotus. *The Histories*, trans. Robin Waterfield and intro. Carolyn Dewald. London: Oxford University Press, 1998.

Herr, Michael. *Dispatches*. New York: Avon Books, 1968.

Hill, Kenneth. *Lost Person Behaviour*. Ottawa: National SAR Secretariat, 1997.

Hill, R. C., and M. Levenhagen. "Metaphors and Mental Models: Sensemaking and Sensegiving in Innovative and Entrepreneurial Activities, *Journal of Management*, vol. 21, no. 6 (1995).

Jamieson, Bruce, and Torsten Geldstezer. *Avalanche Accidents in Canada.* Vol. 4: *1984–1996.* Revelstoke, BC: Canadian Avalanche Association.

Johnson-Laird, P. N., and K. Oatley. "Basic Emotions, Rationality, and Folk Theory," *Cognition and Emotion,* 6, 201–223 (1992).

Kiley, Deborah Scaling, and Meg Noonan. *Untamed Seas.* New York: Houghton Mifflin, 1994.

Kilhstrom, J. F. "The Cognitive Unconscious," *Science,* 237 (1987).

Krakauer, Jon. *Into Thin Air.* New York: Villard Books, 1997.

Kübler-Ross, Elizabeth. *On Death and Dying.* New York: Macmillan Publishing Co., 1969.

Lao-tzu. *Tao Te Ching,* trans. Victor H. Mair. New York: Bantam Books, 1990.

Leach, John. *Survival Psychology.* New York: New York University Press, 1994.

LeDoux, Joseph. *The Emotional Brain: The Mysterious Underpinnings of Emotional Life.* New York: Simon & Schuster, 1996.

———. *The Synaptic Self: How Our Brains Become Who We Are.* New York: Viking, 2002.

Leschak, Peter M. *Ghosts of the Fireground.* New York, HarperCollins, 2002.

Leslie, Edward E. *Desperate Journeys, Abandoned Souls: True Stories of Castaways and Other Survivors.* New York: Houghton Mifflin, 1988.

London, Jack. *Best Short Stories of Jack London.* Garden City, NY: Garden City Books, 1901.

Lopez, Barry. *Arctic Dreams: Imagination and Desire in a Northern Landscape.* New York: Bantam Books, 1986.

Mack, Arien, and Irvin Rock. *Inattentional Blindness.* Cambridge, MA: MIT Press, 1998.

Mayer, Richard. *Thinking, Problem-Solving, Cognition.* New York: W. H. Freeman & Co., 1983.

Newell, A., and H. Simon. *Human Problem Solving.* Boston: Little, Brown, 1972.

Nisbett, R. E., and T. D. Wilson "Telling More Than We Can Know: Verbal Reports on Mental Processes," *Psychological Review,* 84 (1977).

O'Keefe, J. and L. Nadel. *The Hippocampus as a Cognitive Map.* New York: Oxford University Press, 1978.

O'Mara, Shane M. "Spatially Selective Firing Properties of Hippocampal Formation Neurons in Rodents and Primates," *Progress in Neurobiology,* vol. 45 (1995).

Orwell, George. *1984.* New York: Harcourt Brace Jovanovich, 1949.

Perrow, Charles. *Normal Accidents: Living with High-Risk Technologies.* Princeton: Princeton University Press, 1984.

Philbrick, Nathaniel. *In the Heart of the Sea.* New York: Penguin Books, 2000.

Pierpont, Claudia Roth. "Tough Guy," *The New Yorker,* February 11, 2002.

Plato. *Republic,* trans. G. M. A. Grube. Indianapolis: Hackett Publishing Company, 1992.

Read, Piers Paul. *Alive: The Story of the Andes Survivors.* New York: Avon Books, 1975.

Remarque, Erich Maria. *All Quiet on the Western Front,* trans. A. W. Wheen. New York: Ballantine Books, 1982.

Robertson, Dougal. *Survive the Savage Sea.* Dobbs Ferry, NY: Sheridan House, 1994.

Sagan, Scott D. *The Limits of Safety: Organizations, Accidents and Nuclear Weapons.* Princeton: Princeton University Press, 1993.

Saint-Exupéry, Antoine de. *Airman's Odyssey,* trans. Lewis Galantière. New York: Reynal & Hitchcock, 1942.

Selye, Hans. *The Stress of Life.* New York: McGraw-Hill, 1956.

Siebert, Al. *The Survivor Personality.* New York: Penguin Putnam, 1996.

Simpson, Joe. *Touching the Void: The Harrowing First-Person Account of One Man's Miraculous Survival.* New York: HarperCollins, 1988.

Stockdale, James B. *A Vietnam Experience.* Stanford: Hoover Press, 1984.

Suzuki, Shunryu. *Zen Mind, Beginner's Mind,* New York: Weatherhill, 1970.

Tolman, E. C. "Cognitive Maps in Rats and Men," *Psychology Review,* 55 (1948).

Walbridge, Charlie, ed. *River Safety Report.* American Canoe Association, Birmingham, AL: Menasha Ridge Press, 1986–99.

Waldrop, M. Mitchell. *Complexity.* New York: Simon & Schuster, 1992.

Wilde, Gerald J. S. *Target Risk 2: A New Psychology of Safety and Health.* Toronto: PDE Publications, 2001.

Williamson, Jed, ed. *Accidents in North American Mountaineering.* Golden, CO: American Alpine Club, 1997–2001.

Wolfe, Tom. *The Right Stuff.* New York: Bantam Books, 1979.

Yerkes, R. M., and J. B. Dodson. "The Relation of Strength of Stimulus to Rapidity of Habit-Formation," *Journal of Comparative Neurology and Psychology* (1908).

Zuckerman, Marvin. "Sensation Seeking: A Comparative Approach to a Human Trait," *The Behavioral and Brain Sciences,* 7 (1984).

1999 Wilderness Risk Management Conference Proceedings, October 1999, Anchorage, Alaska.

2001 Wilderness Risk Management Conference Proceedings, October 26–28, 2001, Lake Geneva, Wisconsin.

Center for Neural Science, New York University Statement "Overview" Web site.

Outside Online, "Lost & Found: Reports from the field," January 29, 2002.

Trinity College, Dublin, "Mental Models" Web site.

ACKNOWLEDGMENTS

EVERYTHING I'VE EVER written has been a collaboration of sorts. I would like to mention my family, especially my children and parents, who have graciously tolerated silence, weird behavior, and long absence. My daughters, Elena and Amelia, gave selflessly and courageously of their love, support, and insights. John Rasmus provided me with years of patience and guidance; this book grew out of the essays I wrote for *National Geographic Adventure Magazine,* which he founded and where he is now editor. Jim Meigs, executive editor there, shepherded many of my articles along during those years. Jonas and Betsy Dovydenas, Tony Bill and Helen Bartlett, and Eugenie Ross Lemming and Bobby Singer gave me encouragement, friendship, and shelter over the decades. I couldn't have written this book without the help of my editor, John Barstow. And I wouldn't have wanted to write it nearly so much without my agent and friend of thirty years, Gail Hochman. Ari Chaet, Shannon Moffett, and Shane O'Mara read and helped to correct the manuscript. Thanks to my brother Stephen and my son, Jonas. In addition, I am grateful to Debbie every day.

INDEX

AUTHOR'S NOTE

AS IS WELL KNOWN in forensic circles, eyewitness accounts are notoriously faulty. Perceptions can be unreliable, too, especially in times of stress. So there are often disparate versions of what happened during an accident, usually about as many as there are people involved. While I've tried my best to recount the accidents in this book accurately, using official accident reports where possible, there may be other views of some elements of these accounts that are equally valid. The case of my father's crash illustrates the problem well. When we returned to Germany, I interviewed a number of people who were on the scene on January 23, 1945, when his plane came down. Every one of them told me a radically different story of what happened. Similarly, when I interviewed the survivors of the Mount Hood accident in the days after it happened, they too told different stories and remembered events in different ways. (Cliff Ransom assisted in conducting some of the interviews.) So although there may be differences of opinion about the details of some accidents, I have attempted to use what I believe to be the most accurate and reasonable accounts available.

Several names have been changed to protect the privacy of the subjects.

—L.G.